Dodo Destiny

To Chris & Randy,
Here's proof that even Dodos
can write books –
about Dodos....
With best wishes,
Tom

Dodo Destiny

An American Eye on Mauritius

Tom Parker

Dodo Destiny: An American Eye on Mauritius

Published by
Raconteurs Press L.L.C.
Seattle, Washington USA

www.raconteurs.com

Library of Congress Control Number: 2012901888
ISBN 978-0-9719258-3-0 (paper)
ISBN 978-0-9719258-4-7 (Kindle)
ISBN 978-0-9719258-6-1 (ePub)

Printed in the United States of America

Publisher's Cataloging-in-Publication data

Parker, Thomas Reams.

Dodo destiny : an American eye on Mauritius / by Tom Parker.

p. cm.

ISBN 978-0-9719258-3-0 (pbk.)
ISBN 978-0-9719258-4-7 (Kindle)
ISBN 978-0-9719258-6-1 (ePub)

Includes bibliographic references.

1. Mauritius--Description and travel. 2. Mauritius--History. 3. Mauritius--Civilization. 4. Dodo. I. Title.

DT469.M455 P37
969.8/2 –dc23
2012901888

A Gentle Disclaimer

The aim of this book is to provide an introduction to Mauritius, past and present, with special attention to its unique natural history. It is not intended as a replacement for a comprehensive guidebook or an encyclopedia. Maps and illustrations are provided primarily for historical purposes and should not be considered to be accurate. While every effort has been made to ensure that the information in this book is correct and current as of the publication date, errors or inaccuracies may exist. The author and publisher accept no liability or responsibility for errors and omissions, or related damages.

Readers interested in visiting Mauritius and other Indian Ocean islands are advised to consult regional government tourism authorities, airlines, and the ever-growing number of online travel resources to obtain current information and options for planning travel arrangements.

1. An 1844 illustration of dodos and Rodrigues Solitaire

TABLE OF CONTENTS

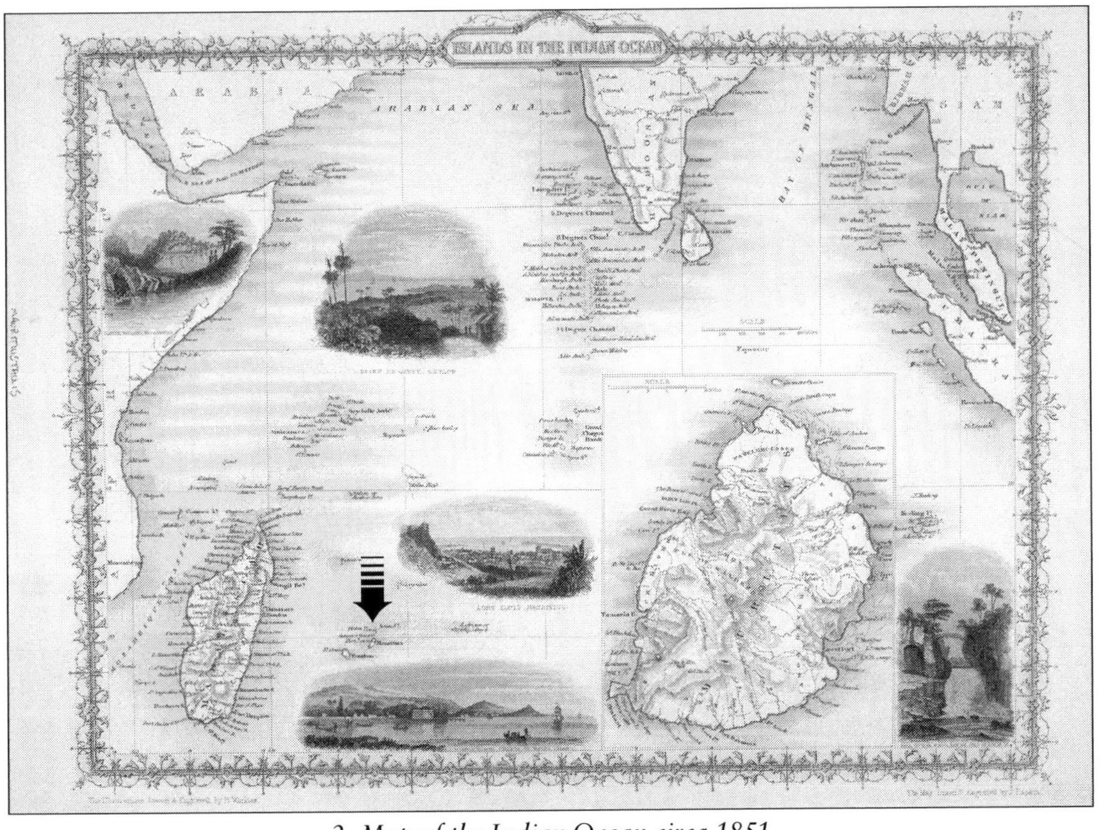

2. *Map of the Indian Ocean circa 1851*

CHAPTER 1

INTRODUCTION

For much of its history, Mauritius was a way station for voyagers traveling between Africa and Asia. And so it was for me in the early twenty-first century. When booking flights between South Africa and Australia as part of a round the world adventure, my travel agent advised me to go through Mauritius to change planes. I knew very little about Mauritius other than it was a popular high-end vacation destination for Europeans. I also had a vague notion that it was connected with the famous but extinct dodo bird.

The island of Mauritius is one of those few tiny dots of land east of Madagascar, floating in the immensity of the Indian Ocean. It is part of an island trio along with islets known as the Mascarenes. This archipelago is a chain of volcanic islands that emerged from the sea around eight million years ago but remained pristine and free from human influence for most of its history. It wasn't until late in the sixteenth century that humans decided to occupy and make their mark on this lush natural laboratory of evolution. Even though the Portuguese bestowed the name of Mascarenes, after navigator Pedro Mascarenhas who visited the islands in 1513, they did not make the tropical islands an outpost.

It remained uninhabited until 1598 when a squadron of Dutch ships en route to Asia was blown off course during a storm and its crew found themselves on Mauritius, a 720-square-mile island almost entirely surrounded by coral reefs. There they found no humans but they did encounter huge numbers of very odd-looking large flightless birds that we know today as the dodo, along with herds of giant tortoises and the largest known species of skinks, a type of lizard.

Once humans decided to call Mauritius and the other Mascarene Islands home, these unique species disappeared. Most of these long forgotten species lasted a little longer than the dodo but left few memories of their existence. Yet the dodo still remains in our collective consciousness. The dodos existed with humans for less than a hundred years, and then they were gone forever, extinct, void. Except in our imaginations.

3. Municipal Theatre, Port Louis, circa 1822

For the next few centuries, Mauritius became a stop for growing numbers of travelers transiting the Indian Ocean and an outpost for European nations trying to build empires of trade. First came the Dutch, then the French, and then the British. Today it is home to a diverse population with Indian, African, and Chinese lineage. All the cultures have left their marks on this island, creating an exotic multilayered cake of ethnicities.

Between the history, unique environment, and the vibrant culture, Mauritius intrigued me. I decided to extend my stopover in Mauritius to four days before continuing on to Australia. Most accommodations for overseas visitors can be found in five-star resort hotels with spectacular white sand beaches. With beautiful scenery, superb restaurants, and all requisite tourist amenities self-contained, many foreigners never leave the confines of their Mauritian resort, which is entirely understandable. However, I wanted to discover more of the "real" Mauritius if only for a few days by parking myself in a central location from which I could explore the island by foot, bus, or taxi. I opted to stay in a large hotel in Port Louis, the capital and largest city.

I didn't know much going into Mauritius. But I did manage to learn that Mauritius was a British colony from 1810 until independence in 1968 and I imagined that I would encounter a strong British influence both in language and customs. I realized that this wasn't necessarily the case when I checked into my hotel, and was greeted by a very friendly staff, all of whom were young, of Indian descent, chicly attired, and speaking French.

The cultural disconnect became clearer the next day when I explored Port Louis by foot. Modern skyscrapers tower over quaint and sometimes dilapidated French colonial buildings that share streets with high-rise apartment blocks. The high-density, compact urban city contrasts starkly with many square miles of sugarcane fields that cover much of the island. Once outside the city, much of the island feels almost empty and sparsely populated. Port Louis is the only city resembling an urban metropolis between Johannesburg and Mumbai, which is surprising for a place that appears only as a tiny speck of land on a map.

The juxtaposition of cultures can be jarring. A stern-looking statue of Queen Victoria competes for attention with a nearby statue of Mahé de Labourdonnais, a notable governor of the island during eighteenth-century French rule. Port Louis has its own Chinatown, a large Catholic cathedral, a huge Islamic mosque, and numerous Hindu and Tamil temples. During the day, Port Louis seemed like an Indian city with crowded streets filled with many women wearing colorful saris. Everyone seemed to be speaking in French. Later I discovered they were actually speaking Mauritian Creole. The lingua franca of the country, Kreol, borrows heavily from spoken French.

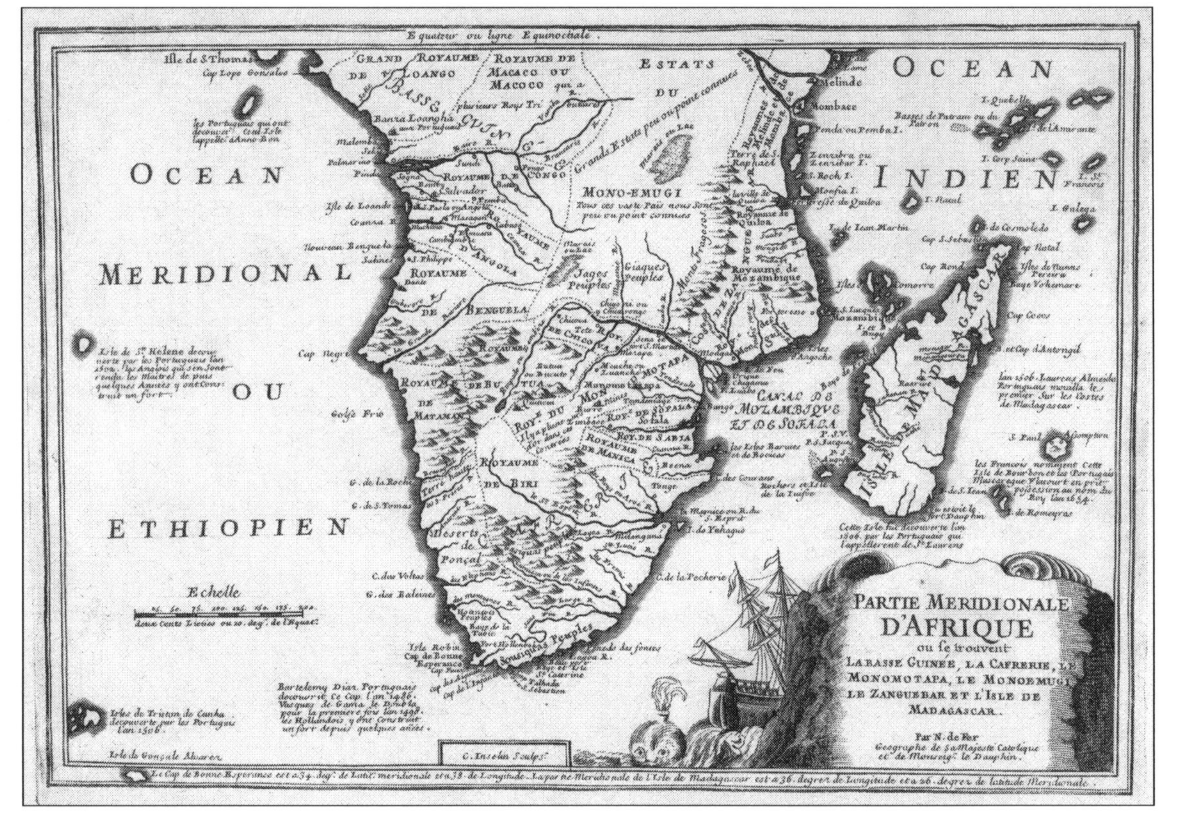

4. French map of southern Africa circa 1700

My cultural confusion increased as I noticed the unusual mix of street names. Many seemed to be named after historically significant people, not necessarily from Mauritius, as President Kennedy Street would indicate. Other street names included Lord Kitchener, Mere Barthelemy, Sir William Newton, Jummah Mosque, Rue de la Reine, Sir Seewoosagur Ramgoolam, Farquar, Imam Moustapha, and Abattoir Road. Clearly I wasn't in Kansas.

As a first-time visitor, I couldn't help but notice that the dodo is everywhere. Gift shops are filled with toy dodos, dodo-themed beach towels, dodo handbags, and all sorts of merchandise festooned with dodo images. Many stores and restaurants include the name dodo, and a picture of the bird appears on the official government coat of arms. The dodo was blasted into my consciousness but with little explanation as to exactly what this bird was. Clearly there was more to its historical significance than a cartoon character on a beach towel might suggest. The more dodo images I saw, the more I wanted to learn about this strange looking bird. I knew the phrase "Dead as a Dodo" from childhood, but now for the first time wanted to know how and why it became extinct.

During my four-day visit, I did my best to see the major sites but there wasn't enough time.

The strange disconnect between different cultures living in close proximity, the wide range of contrasting architectural styles, the large number of people of South Asian descent speaking French in a former British colony where the official language is English—it all seemed odd to me but in a very appealing way.

With a history of European settlement dating to the seventeenth century in a very remote corner of the world, Mauritius seemed very old but at the same time new, like a snapshot of the future. I found myself wondering why it was that I knew so little about this place and was determined to return at some point for a longer visit.

TIMELINE

26 MILLION YEARS AGO

Ancestors of Nicobar Pigeon begin to island hop across Indian Ocean

8 MILLION YEARS AGO

The volcanic island of Mauritius emerges from the sea

5 MILLION YEARS AGO

Reunion Island emerges

1 TO 3 MILLION YEARS AGO

Rodrigues island emerges

1498

Portuguese explorers stumble upon Mauritius in the wake of Vasco da Gama's voyage around the Cape of Good Hope.

1510

Portuguese navigator Pedro Mascarenhas visits the island and names it Ilho do Cerne, or Cirné. The Portuguese use it as a way station for food and fresh water but do not establish a permanent settlement.

1528

Portuguese discover Rodrigues Island.

1598

The first Dutch expedition arrives and claims the uninhabited island. They rename it after their head of state, Maurice, Prince of Orange and Count of Nassau.

1598–1620

Living dodos are exported to Europe and soon appear in paintings and print descriptions.

1630

Sightings of dodos on Mauritius become very rare.

1646

France occupies Île Bourbon (today called La Réunion)

1638–58

Dutch settlement begins and ends in failure, island is abandoned.

1664–1710

Second Dutch attempt at colonization fails. The Dutch withdraw permanently.

1682

Last documented sighting of a dodo in the wild on Mauritius.

1690

The approximate date by which the
Mauritius dodo is believed to be extinct.

1715

French East India Company claims Mauritius
for France, renames it Île de France.

1721–67

Settlement begins; Port Louis founded as a
base for attacking the British in India.

1744

St. Geran tragedy occurs, inspiring the story of
Paul et Virginie that is published in 1787.

1767

French East India Company sells Mauritius; control
of island transferred to French crown. Pierre Poivre
arrives as Intendant for Île de France and Île Bourbon.

1768

Poivre creates a botanic garden at
Mont Plaisir in Pamplemousses.

1770

Nutmeg and clove trees successfully purloined
from Dutch-controlled Spice Islands and
planted at Pamplemousses, breaking the
Dutch monopoly on the spice trade.

1787

Jacques-Henri Bernadin de Saint-Pierre publishes *Paul et Virginie* which becomes a best seller in Europe.

1803

General Decaen, the last French governor of Île de France is appointed.

1810

Battle of Grand Port: France defeats Britain in its only naval victory of the Napoleonic era. Later that year, overwhelming British forces invade and capture Mauritius. General Decaen surrenders.

1812

Colonel Draper founds the Mauritius Turf Club, which opens the first racecourse in the southern hemisphere and is today is one of the oldest horseracing authorities in the world.

1814

Treaty of Paris. Mauritius, Seychelles, and Rodrigues ceded to Britain, La Réunion is returned to French rule.

1834

British abolish slavery. It is phased out on the island under a transition period known as apprenticeship.

1835

Indentured labor system introduced. In subsequent decades hundreds of thousands of workers arrive from India.

1847

The two-pence "Post Office" Blue Mauritius postage stamp is issued and later becomes one of the rarest and most valuable stamps in the world.

1848

The Dodo and Its Kindred is published, written by Hugh E. Strickland and Alexander G. Melville and asserts that the dodo became extinct due to human activity.

1859

Charles Darwin's *Origin of Species* published.

1865

British schoolteacher George Clark discovers cache of dodo bones on Mauritius which are sent to the British Museum for analysis.

1865

Alice's Adventures in Wonderland by Lewis Carroll is published and includes an illustration of a dodo by John Tenniel.

1866–68

A malaria epidemic kills more than 40,000 people and causes a mass relocation from Port Louis to higher elevation communities in the central plateau area of Mauritius.

1924

Indentured labor system is formally abolished.

1926

First Indo-Mauritians elected to government council.

1968

Mauritius becomes an independent country.

1992

Mauritius becomes a republic within the British Commonwealth.

CHAPTER 2

RETURN TO MAURITIUS

It took me several years but thanks to unabated curiosity, I returned to Mauritius for a closer look. My initial visit sparked an undimmed interest in all things Mauritian that I came across in books and news reports. This time around I decided that a stay of several months was in order. I also learned much more about the nearby French island of La Réunion, and wanted to see this island that shares so much of its history, culture, and language with Mauritius. During the long period of French rule, Mauritius was called Île de France while prior to the French Revolution, La Réunion was known as Île Bourbon.

Mauritius has carefully cultivated an image as a high-end tourist destination. After independence from the UK in 1968, the new Mauritian government sought an economic development strategy to diversify from its traditional economy that was almost entirely dependent on sugarcane, a single crop susceptible to periodic disease, storms, and market price collapse. Several industries were identified as prospects including tourism. With its stunning white sand beaches, beautiful scenery, and a plentiful supply of potential employees, the hotel industry was a natural choice for Mauritius and it has become an economic success story.

Mauritius is a small island, however, with a finite number of beaches suitable for resort development. Consequently the decision was made to limit hotel development to only the high-end tourist segment and focus on luxury hotels. Mass charter flights were banned, a number of world-class resort hotels were built and carefully crafted marketing campaigns attracted growing numbers of overseas visitors. Today Mauritius receives nearly one million tourists annually from many countries, nearly seventy percent from Europe with the majority coming from France. It's worth noting that Americans make up only a tiny percentage of all tourist arrivals, less than 7,000 in 2010. I think this helps explain why Mauritius remains unknown to most Americans.

Despite its reputation for extravagant and expensive resort hotels, Mauritius offers more affordable

5. View of Port Louis from the sea in 1844

accommodations. These include small hotels, villas, and self-catering apartments for independently minded travelers on a budget. Many of these apartments can be found in Grand Baie, a town north of Port Louis named for its beautiful tranquil harbor that is close to a number of beautiful beaches. Another popular spot with apartment rentals is the town of Flic en Flac on the southwest coast, famous for having the longest beach in Mauritius. After reviewing a few possibilities, I opted for an apartment in Grand Baie for a prolonged visit.

Grand Baie ("bay" in English) was once a small and charming fishing village and today has grown into the center of mass tourism for Mauritius, for better or worse. The bay is still beautiful with clear turquoise waters that allow a huge number of activities including sailing, diving, and snorkeling. The town center of Grand Baie, however, is crowded with many shops, noisy nightclubs, restaurants, and all manner of tour operators. Proliferating real estate development projects throughout the area contribute to an image of a burgeoning tourist center.

I wouldn't call Grand Baie a tourist trap exactly, and compared to similar-sized seaside tourist towns I have seen in the U.S. and other countries, it's quite pleasant. Away from the center of town, there are many quiet residential streets particularly in the Point aux Cannonieres section where elegant homes hide behind tall security gates and imposing privacy walls. The beautiful (quite

6. Grand Baie

stunning, in fact) public beaches of Trou aux Biches and Pereybere are both only a short drive from Grand Baie.

In addition to a plentiful supply of self-catering apartments, Grand Baie offers other advantages for the overseas visitor planning an extended stay. There is much more nightlife here than in Port Louis, and it is home to a large number of restaurants and stores. This includes the large Super-U hypermarche; a French-owned superstore that sells everything from groceries, wine, clothing, and hardware to incense and festive decorations for important Hindu holidays.

Grand Baie is a convenient transportation hub with bus and taxi service connecting to all cities on the island. A particularly useful feature of Grand Baie for me was the air-conditioned express bus service to Port Louis that leaves every thirty minutes every day. The fare for this trip was a tantalizing thirty rupees, about one U.S. dollar. The scenic drive through small towns, vast sugarcane fields, and into the urban neighborhoods of Port Louis was worth every rupee.

Driving in Mauritius leaves a lot to be desired and can be frightening at times. There are thousands of serious road accidents each year and an alarming number of these are fatal. Mauritians are very polite and friendly people but I get the impression that once behind the wheel of a car, some of them, particularly bus drivers, can act like frustrated Formula One race car drivers. Traffic in Port Louis can be very chaotic, especially during rush hours.

Mauritius uses the British system of driving on the left side of the road and many cars are much smaller than those found in the U.S. There is only one large highway in the country, connecting the international airport and Port Louis. Many other routes are winding two-lane roads in various states of repair that are often crowded with cars, trucks, buses, and motorcycles constantly trying to overtake each other. Distracted drivers talking on mobile phones only adds to the fear factor.

It's a common sight to see motorcycles dangerously weaving in and out of heavy traffic, and drivers of both cars and trucks will try to pass one another near blind intersections and other risky spots. As a bus passenger gazing out the window, I've witnessed several horrible accidents, which has not inspired me to drive myself in rental cars. Taxis and the public bus system are an easy alternative to driving a rental car, particularly for those of us accustomed to cars with steering wheels on the left.

The public bus system in Mauritius was a mystery to me. It is a service operated by a number of private companies serving different geographic regions. Published timetables were difficult to decipher and turned out to have only a wishful correlation to actual travel times. Traveling anywhere between the north and south by bus in Mauritius requires a transfer in Port Louis at two separate bus stations that are some distance from each other. While the public bus systems are privately owned and operated, the government subsidizes fares for students and senior citizens who ride for free. When school lets out in the afternoon, hundreds of neatly uniformed schoolchildren can be seen lined up at bus stops, ready to swarm into the next available bus. Chaos ensues as they scramble for seats.

Friendly service and inexpensive fares compensate for whatever the buses may lack in sophistication and for the sometimes alarming driving techniques. Buses are very colorful and remind me of buses in India, with festive Hindu decorations plastered around the driver. Many times I couldn't figure out which bus to take but the driver, conductor, and often friendly fellow passengers unfailingly helped me out every single time. While

7. A seaside Hindu shrine in Grand Baie

bus travel isn't for everyone, I found every trip to be an adventure. A number of times, I couldn't help but notice that I was the only passenger not wearing a brilliantly colored sari while speaking Kreol.

Traveling around Mauritius by public bus isn't quick or even easy, as buses can be hot and crowded and are frequently trapped in traffic jams, particularly around Port Louis. However, travel by public bus has its rewards. It offers visitors something they will never see from the confines of their resort hotels, a window into the lives of the vast majority of people who live in Mauritius.

Taxis are plentiful on Mauritius and relatively cheap but they are not metered, and bartering to settle on a fare before accepting a ride is highly advisable. If you look like a foreigner out for a stroll on a public street, be prepared to have every passing taxi driver ask you if you want a ride. Usually this is just a nuisance but I've run across aggressive taxi drivers competing for fares that can make for an unpleasant experience. Taxis are convenient, however, and it's customary to hire one for a half-day or whole-day tour. It is a great way to explore the island.

My apartment in Grand Baie was situated in a densely populated, mostly Hindu neighborhood about 30 minutes walking distance from the center of town. My landlords, a gracious Hindu couple, did their best to make me feel welcome. They invited me to family events, including the festival of Diwali, which is somewhat akin to being invited to a family Christmas dinner party.

My prolonged stay in Grand Baie was an intriguing introduction to the Hindu influence on Mauritius, a group that today represents the majority of the population and holds most of the national political power.

Mauritius has no indigenous population and the different ethnic groups comprising today's population resulted from successive waves of immigration from around the world. Mauritius is a fragmented society with people descended from Africans, Europeans, South Asians, Malagasies, Chinese, and various combinations of these groups. In 2010, the total population of the country was approximately 1.3 million, making Mauritius one of the most densely populated countries in the world. Yet except for Port Louis and some of the larger towns, it doesn't feel overly crowded, as there are still vast expanses of sugarcane fields, national parks, and mountainous forests.

Indo-Mauritians form the single largest group, including both Hindus and Muslims. The single largest group, Hindus make up half the total population and dominate the public sector. This is a relatively recent development, since prior to independence in 1968 Hindus were largely excluded from the political process.

Today Hindu influence is very visible in domestic architecture. Red flags and many statues of Hindu deities can be seen everywhere. Many if not most Hindu houses have a small outdoor shrine dedicated to their family's favorite Hindu deity, which is often elaborately decorated and surrounded by incense and candles. Small indoor shrines are equally commonplace and are elaborately decorated for holidays. Also scattered throughout the country are public outdoor funeral pyres that can be reserved by families for performing traditional Hindu funeral services of their loved ones.

Muslims represent about 17 percent of the population and are distributed throughout the country. There is usually at least one mosque in many of the smaller towns. Creoles are descendants of African and Malagasy (from Madagascar) slaves and make up about 27 percent of the population, and Sino-Mauritians of Chinese descent comprise about three percent. Franco

Mauritians form less than two percent of the population, and are descended from French colonists. While they are only a tiny segment of the overall population, Franco-Mauritians remain quite influential and many retain significant property holdings. British and South African residents are included among expatriate groups living throughout Mauritius. There are also other groups from India and elsewhere including Tamils, Telegus, and Marathis. Intermarriage between different groups has made Mauritius a veritable ethnic melting pot of different races and cultures.

Once ensconced in my new abode, I became familiar with the daily routines of the neighborhood. Every morning I awoke to the calls of the muezzin from the local mosque located on the main road ringing the bay. You can't miss it when walking by—it's directly across the street from the French Catholic Church and not far from a large Tamil temple. Closer to my apartment was a Hindu temple that seemed to be the site of many festive, noisy rituals and celebrations.

My landlords, Avinash and Sonia, couldn't have been nicer, as were their two charming and polite children, a girl age seven and a boy of fourteen. I think I was their first American tenant or at least the first one they had seen in a very long time, which certainly piqued their curiosity.

Adding to this cultural kaleidoscope was the interesting assortment of neighbors in my eight-unit apartment building. My third-floor unit had a pleasant balcony that overlooked sugarcane fields and volcanic mountains in the distance. An eccentric middle-aged Italian woman and her very elderly father occupied the apartment directly across the hall from mine.

Neither of them spoke any English and we communicated in French, or more accurately, I attempted to do so, speaking my poor version of French. My neighbor was very pleasant if loud, and on most mornings while having coffee on her balcony, she would burst into song, usually with romantic Italian arias. Mamma mia. The mood next door usually deteriorated at least once in the afternoon with a heated and emotional argument with her father in angry outbursts of Italian expletives, but always seemed to cool off by evening. The other neighbors were mostly French or other Europeans who rented the apartments on a short-term basis and were mostly quieter.

From my balcony overlooking the street I could look down and see local children riding their bicycles while their mothers chatted with each other, wearing saris and other traditional Indian garments. Once a day, a vegetable vendor would stop his truck in front of my building, which drew a crowd from many blocks, mostly women buying essentials for dinner. With my neighbors conversing in Italian on the balcony next door, combined with occasional French conversation wafting down from the apartments upstairs, I felt like I was living in a global village.

The multicultural aspect of Mauritian life became much more vivid to me from watching the nightly news on local television. The Mauritian Broadcasting Corporation (MBC) is the national network that broadcasts the same 30 minute evening news broadcast sequentially in three languages: French, English, and Hindi. The contents for all segments seem to be identical in terms of topics and video footage but are presented by different news anchors fluent in their respective languages.

To North American ears, the nightly local news broadcast in three very different languages makes Mauritius an exotic and polyglot place. It's actually

8. A beach near the center of Grand Baie

more exotic than that, however, as French, English, and Hindi represent only three of twelve or more languages spoken in the country. Others include Bhojpuri and Urdu. French is the language of the media, used both for newspapers and television broadcasting while Mauritian Kreol is spoken by virtually everyone at home as a primary language.

The dominance of Kreol and French in daily life means English is less commonly spoken, even though it is the official language. Plenty of Mauritians speak English but often with limited proficiency, and many times I've found it necessary to try and communicate with my unfortunate rendition of French. I suppose that with so many languages spoken in a country of just 1.3 million, Mauritius is an extreme case of multilingualism. This can be confusing for English speaking visitors but I find it an aspect of Mauritian life that makes it such a unique and appealing place.

The programming also includes local television production from very low budget children's shows to Creole language soap operas. While perhaps lacking in production values, and spoken in what was for me a mysterious language, I found these Creole soap operas to be a fascinating insight into Mauritian life. Bollywood films and television shows imported from India make up a large part of evening programming. Most local Mauritius broadcast television content seemed very conservative and subdued compared to what passes as entertainment on American television. Cooking shows are popular and usually feature a matronly woman, wearing a very formal sari, preparing and explaining traditional Indian recipes in Hindi using a spotless kitchen that bears no resemblance to my own.

The MBC does not have the financial resources of larger international news organizations for reporting, but makes a notable effort to report on all aspects of Mauritian life and to publicly recognize contributions made by different religious and ethnic groups. The daily television news includes local stories about many small communities and festivals for every major religion observed in the country. I also watched frequent subtle public service announcements that encourage viewers to respect their neighbors' different religious practices and celebrate their festivals, many of which are national holidays. I'm not sure what the local response is to these sermon-like messages, but in the context of Mauritian life, I can understand the motivation for trying to inject a little socially cohesive glue like this through the medium of television.

The realities of having a relatively large population confined to a small island means everyone needs to get along with each other out of necessity. There are certainly undercurrents of ethnic tension in Mauritius and there have been open conflicts in the past. Potential flash points for unrest include friction between Hindu and Muslim groups, and resentment by mostly Catholic Creoles of the Hindu-dominated government.

Mostly, however, everyone lives in harmony, which is a big factor contributing to the country's economic success. Divisions between different religious groups and ethnic communities are quite distinct in Mauritius but I also noticed a very strong common sense of national identity. People are genuinely proud to be Mauritian whatever their ethnic or religious background. This pride is evident in the energy that residents have put into their communities and economy.

For example, Grand Baie has grown rapidly in the past thirty years, raising the economic fortunes of the local population. Many families formerly dependent on meager livelihoods in sugarcane cultivation have seen their personal fortunes rise as they have embraced opportunities in the burgeoning tourist industry over the past thirty years. They may not be rich by Western standards but they are much more prosperous than their parents' generation and have high hopes for their own children. There seem to be quite a few entrepreneurial families in Grand Baie who operate small tourist-oriented businesses ranging from sailboat charters to hotels and restaurants.

I met one such family while taking a stroll one morning to Grand Baie along the Route Royal, the main road that rings Grand Baie. I stopped to take a look at a small Hindu shrine mounted on a pedestal in the sea when a young man walking with his cute one-year old daughter noticed my interest. He told me that it was a shrine to a goddess of the sea. He introduced himself as Vinay and said that his family owned the apartment building next door that had wonderful tourist apartments for rent and would I like to see it? While clearly "enthusiastic" he didn't seem pushy and I agreed. The apartments were in fact extremely nice and well furnished. I told him I already had accommodations but would keep it in mind for future reference.

Whereupon, Vinay said that in addition to apartment rentals, his family operates a catamaran sailboat tour company, a small rental car business, and they have a Tandoori Indian restaurant directly across the street for which he gave me a 10% discount card with his personal signature. All in all, it was quite a sales pitch for 8 o'clock in the morning but fairly typical for small tourist-oriented businesses in Mauritius.

My extended stay in Grand Baie was intended as an immersion experience in Mauritian culture.

While not entirely unexpected, I was surprised to the extent it became an immersion in Hindu culture. This was partly a result of living in a mostly Hindu neighborhood but a pleasant surprise nonetheless. Having settled in at Grand Baie, I turned my attention to exploring the rest of Mauritius and particularly Port Louis.

CHAPTER 3

MEET ME IN PORT LOUIS

PORT LOUIS IS ONE OF THE MOST INTRIGUING CITIES that I've come across. It's hard to categorize because it reflects so many different cultures and architectural styles, all in a relatively confined space. Port Louis is not only the largest city in Mauritius with a population of around 150,000, but it is also the capital of the country, plus a growing economic hub for the entire Indian Ocean region. The influx of commuting office workers from towns all over the island swells the weekday population of the city, making Port Louis seem much larger than it really is.

It is also the transportation crossroads of Mauritius or maybe more accurately described as the center of

9. Caudan Waterfront, Port Louis

gridlock during rush hours. The number of vehicles on the nation's roads increases by several thousand each year and the expansion of the road system hasn't kept pace. Traffic in and out of Port Louis at rush hour can be severely congested, a situation that is becoming worse each year. Several major road improvement projects are underway and should help remedy the situation when they are completed. A bypass is being built to divert through-traffic from the city streets of Port Louis to reduce congestion.

A major focal point of Port Louis is the Caudan Waterfront complex. Built alongside the harbor in 1996, this modern shopping center has offices, hotels, restaurants, a cinema, bank, a small museum, casino, and more than 150 shops. It is connected to the rest of Port Louis via two large pedestrian tunnels that lie directly under the busy motorway that bisects the city. Police and security guards are a highly visible presence throughout the complex, adding a reassuring note for foreigners shopping in the pricey duty-free jewelry and other stores.

The Caudan Waterfront is designed specifically for overseas tourists and isn't representative of the rest of Mauritius. That said, it's a nice place to visit and offers a calm oasis compared to the urban chaos in the rest of the city. The Caudan has two excellent hotels, more than 160 shops, a cinema, cafes, bars, and restaurants offering alfresco dining at tables overlooking the harbor. In many countries, the capital city is also a national center for nightlife and entertainment. Not in Port Louis. The Caudan is the only place where nightlife can be found. That makes it a popular spot both with tourists and locals.

The architecture of the Caudan complex is a bit puzzling, sort of contemporary tropical that incorporates a few elements of traditional colonial architecture. There are also a few theme park touches like the giant gold lion statue sitting on top of the entrance to the casino. It was only after many visits that I took notice of how every building is very solidly built of stone or concrete with heavy storm doors and shutters to protect windows and

10. *View of Port Louis from the "Pouce" mountain in 1858.*

doors against cyclones. Cyclones are powerful storms bringing torrential rains and peak winds exceeding 250 km/hr. that can flatten everything in their path. Mauritius is located in the middle of an active cyclone region and while direct hits are rare, they have occurred regularly over the centuries causing great damage and destruction to all of the Mascarene Islands.

This should not have surprised me because tropical heat brings tropical storms. Port Louis is situated in a narrow valley with the harbor to the west and semi-circular barrier of mountains to the east. The valley traps the heat baking Port Louis, consequently temperatures are hotter than the rest of Mauritius, especially during summer. The French chose Port Louis for the capital because of the fine natural harbor and the relative lack of wind, which protects ships at anchor. However, lack of wind also means more heat, making Port Louis one of the warmer spots on the island.

Since it is on the other side of the equator, the seasons are reversed from the Northern Hemisphere. No matter what time of year, be prepared for warmth. Mauritius has a tropical climate with only two seasons. From November to April, summer is hot and wet. The winter is warm and mostly dry, lasting from May to October. The coasts are hotter than the central plateau in the interior of the island. Summer temperatures on the coast start out comfortably at dawn at around 24C (75F) but can reach 30C (86F) by noon. Breezes from the harbor help cool the Caudan Waterfront but the rest of Port Louis can get a bit uncomfortably hot during summer. Most buildings are air-conditioned and the heat and humidity seem more bearable to me than in many U.S. cities like Washington, D.C. in summer.

The cyclone season lasts from January through March, and while most storms bring nothing more than heavy rains and strong winds, the threat of serious damage is very real. Mauritius has a very sophisticated and advanced system of cyclone warnings and emergency procedures in place that range from preliminary warnings to orders to stay indoors. There are public emergency shelters located throughout the island designed specifically for cyclones.

I don't know the specifics of the local architectural codes but I have been told that all new buildings in Mauritius must be designed and constructed to withstand cyclones. Once a cyclone hits, it can easily take more than a week for electricity and other utilities to be restored. Most hotels and large office buildings have emergency back-up generators and water supplies, as do many apartment buildings and private homes.

When the subject comes up, I've asked Mauritians what it is like to experience a severe cyclone. My landlords in Grand Baie brushed them off as nothing serious, even calling them fun, just an opportunity to stay indoors for a couple of days with their family. I'm not sure how much fun that would be for me, trapped inside a building with 200 kilometer per hour winds howling overhead.

The economics of building cyclone-proof homes means that most modern single-family houses are built of ubiquitous concrete blocks and are lacking in creative design flair. Driving around small towns throughout Mauritius, I have seen many concrete-block houses sitting unfinished and many remain that way for years. It's common for families to do the construction themselves which is halted when they run out of money. Concrete doesn't deteriorate so they can start building again whenever they find the time and money.

It's only a five minute walk through either of the two large pedestrian tunnels under the motorway that links the orderly, controlled, and somewhat artificial world of the Caudan Waterfront with the rest of Port Louis, which is vibrant, loud, crowded, chaotic, and utterly fascinating. While many buildings are new, particularly the modern high-rises that dominate the skyline, Port Louis is also home to the island's greatest concentration of historic architecture. Many colonial buildings are well preserved while others are in an advanced state of decay.

A malaria epidemic during the 1860s decimated the population of Port Louis, prompting many to relocate to the higher and cooler areas of the central plateau less likely to breed mosquitoes. The majority of Mauritians still live in the central plateau, a succession of towns

11. Statue of Queen Victoria at Government House, Port Louis

extending from Port Louis that include Rose Hill, Quatre Bornes, Vacoas, and Curepipe. Early colonial structures remain in Port Louis today although private houses from this period are rare, having succumbed to cyclones, termites, and neglect.

The architecture of Port Louis can seem British, French, Indian, Chinese, Islamic, or modern American depending on what street you are on and in which direction you happen to look. Port Louis is compact enough for an easy walking tour of the downtown area if you don't mind navigating between busy and sometimes aggressive traffic. Sidewalks are raised quite high above street level with very steep gutters to accommodate torrential rainfall that occurs during the wet season.

The Place des Armes is a major thoroughfare that connects the waterfront to Government House and the Mauritius Parliament in the center of old Port Louis. This grand boulevard is lined with two rows of giant palm trees that overlook a series of historic statues. It's a busy area frequented by thousands of workers crossing streets en route to office towers and is also a major artery for motor traffic. The Place des Armes is more than just a street, however. It's filled with park benches under the palm trees that offer a respite from the sun in a crowded city. A giant statue of the French colonial governor Mahé de Labourdonnais marks the western end near the harbor. The Place des Armes ends at the gated courtyard of Government House, which contains a large statue of a somber-looking Queen Victoria. The route between these two landmarks incorporates statues of other important dignitaries.

Government House dates from around 1740 and the historic French colonial building served as the official residence of Mahé de Labourdonnais, who was probably the most important governor of Île de France (as Mauritius was then known). Labourdonnais was influential in the history of Mauritius and many institutions and places are named in his honor, including the elegant Labourdonnais Hotel in the Caudan Waterfront. Mahé de Labourdonnais transformed Port Louis from a backward outpost into a thriving capital city and undertook major public works projects that cast Port Louis as a major port of call for

12. Statue of Mahé de Labourdonnais, Port Louis

sailing ships from around the world en route to India and Asia.

He was a sailor from an early age, had traveled extensively throughout the Indian Ocean, and first visited Île de France in 1723. He was aware of the strategic value the island had for France, with natural advantages over Île Bourbon (La Réunion) containing a harbor at Port Louis that could become an important naval base and a site for storehouses for French trading with India.

He arrived as governor in 1836, directed by the French East India Company with instructions to administer Port Louis as a trading post and relay station for ships traveling to the east and nothing more.

Labourdonnais was a visionary who saw the potential of Mauritius as more than merely a relay station for ships traveling east. He envisioned the island as a strategic hub that could become a thriving French colony. His vision was matched with his gifts of unusual administrative ability and, by many accounts, he also had enormous physical energy and personally supervised the building of many important public works projects. Labourdonnais provided Port Louis with roads, an aqueduct for drinking water, a hospital, brick-making factories, offices, and first sugar factory. The Dutch introduced sugarcane to Mauritius which became neglected under early French rule. Labourdonnais revitalized sugarcane cultivation and also started exports of locally grown cotton and indigo.

He deepened the harbor, and built new quays and a shipyard for repairs. Labourdonnais is remembered mainly for transforming Port Louis into an essential port of call for all ships traveling to India and Asia. It remained an important port with increasing amounts of ship traffic that lasted until the opening of the Suez Canal in 1867. The opening of the Suez Canal provided a much shorter route from Europe to Asia and resulted in a gradual decline of ships calling at Port Louis.

His official former residence, Government House, is not open to the public so the closest I've been able to get is to gaze through the iron gates that surround the courtyard in front of the building. On a recent visit, the entire building was obscured due to a major renovation project that should restore the building to its former glory.

Located directly behind Government House is the modern Mauritian parliament building where meetings of the National Assembly are held. The government of Mauritius mostly follows the British model of parliamentary democracy with 62 members elected by universal suffrage every five years. The country is divided into twenty constituencies, each of which sends three members to parliament. Rodrigues Island is its own constituency and sends two members.

Mauritius became an independent country in 1968 within the British Commonwealth with Queen Elizabeth II as head of state represented by a Governor General. In 1992, Mauritius became a republic within the British Commonwealth, eliminating the British monarch as head of state and instituting the position of president. The post of president in Mauritius is largely ceremonial. The political party that wins the majority of seats in parliament forms the government and usually its leader becomes the prime minister who wields executive power. Since independence from Britain in 1968, Mauritius has had a history of free and fair elections with a positive human rights record, which is quite an achievement in this sub-Saharan region of the world.

Two blocks from Government House is the Port Louis Market, a wrought iron building dating from Victorian times. Above the entrances to the wrought

13. Jummah Mosque, Port Louis

iron gates is a royal crest with the letters VR for Victoria Regina (Queen Victoria). The market is usually teeming with activity, and if you stand out as a foreigner, vendors will try to sell you any number of things, be it fresh fruit, caged pet birds, straw hats, T-shirts, you name it. If any of the merchandise on sale looks fake, it probably is —this is true for vendors outside of the market as well.

The crowds can be overwhelming but the visual spectacle of the market is entertaining, and very colorful, especially the enormous selection of fresh produce. Huge piles of gorgeous tropical fruits and vegetables are stacked on giant tables. Many are familiar including tomatoes, eggplant, pineapples, and bananas, but other vegetables were completely new to me including "chou chous." This is a green fleshy vegetable that is served boiled as a side dish and is quite good.

The market offers myriad items for sale, including any number of medicinal remedies for whatever ails you. One vendor, a Mr. Mootoosamy, advertises "Tisanes Composees pour Maigrir, Cellulite, Aphrodisiaque, Foie" which translates to " Herbal Teas for Weight Loss, Cellulite, Aphrodisiacs, Stomach Upset. " I think he does a brisk business, as the last time I passed his stall, four young German women were eagerly pushing Mauritian rupee banknotes into his hands. Hope springs eternal, even in this remote corner of the Indian Ocean.

The sidewalks throughout the downtown commercial part of town are lined with street vendors and hawkers selling everything from fresh fruit juices and clothing to sunglasses, handbags, electronics, and music CDs of questionable authenticity. There aren't any large department stores in Port Louis, only a few street-level urban malls that have a collection of 10 to 20 small interconnecting shops.

West of the market, walking along Royal Street towards Chinatown, there are dozens of small storefronts selling all manner of merchandise. A garment district spreads over several blocks highlighted with small clothing and apparel fabric stores, some with rather startling mannequins in the shop windows. I don't think I've seen so many bolts of brilliantly colored fabrics in one place or realized that there is no such thing as having too many sequins.

The atmosphere becomes much more Asian walking further west towards the entrance of Chinatown. The Chinese community in Mauritius is quite old and well established but the Port Louis Chinatown isn't geared towards tourism.

One notable feature of Port Louis is the complete lack of familiar convenience stores or any large stores at all. Outside of the Caudan Waterfront, there is only a single McDonald's that intrudes as America's contribution to modern culture. Most stores are quite small and specialized, which strikes me as a time warp coming from the U.S. In order to buy envelopes, I had to go to a tiny, darkly lit stationery store near Chinatown where the proprietress showed me a selection of envelopes pulled from a dusty glass cabinet. It was a far cry from having my purchase electronically scanned in a brightly lit, oversized OfficeMax.

At the corner of Royal Street and Jummah Mosque streets is, surprise, the Jummah Mosque with its giant minaret towering over other buildings. This vast structure covers an entire block and depending on what time of day it is, you can hear the muezzin call the faithful to prayer.

East and uphill along Jummah Mosque Street is the Citadel. It's a bit of a hike to reach this nineteenth-century fort but it has a fantastic view of all of Port Louis, the harbor, and mountains. Built by the British in 1835, this fort was named Fort Adelaide after Queen Adelaide, the wife of William IV and namesake for the city in South Australia. The fort itself is a bit depressing, consisting of a dark stone structure surrounding a large barracks square.

It was built by the British to protect themselves both from outside attack and internal threats, including French settlers angry about the abolition of slavery that accompanied the British conquest in 1810. Former French sugar plantation owners fiercely resisted efforts to free their slaves until the 1830s and insisted on receiving compensation from the British government before acquiescing.

Looking east from the Citadel is an excellent view of the Champ de Mars racetrack at the eastern end of the Pouce Valley. Under French rule this was a military training ground and the scene of many duels. When the British conquered the island in 1810, the French-speaking residents of Mauritius far outnumbered the resident British administrators and soldiers. An enterprising British Army Colonel, one Edward Alured Draper, had the idea of introducing horse racing at Champ de Mars, in the hopes of fostering good relations with the French populace and diffusing long-standing rivalries.

The Mauritius Turf Club was established in 1812 and the first races were held at Champ de Mars that same year. Horse racing and especially gambling proved to be a huge success. Mauritians are obsessive horse racing fans and keen gamblers. The grandstands at Champ de Mars usually fill to capacity every Saturday during the racing season that lasts from April to November.

Horse racing at Champ de Mars is a highlight of social life in Mauritius and the Mauritius Turf Club has always attracted a distinguished and cosmopolitan crowd. The Duke of York Cup race is a perennial favorite event that attracts thousands. Over the years, visitors have included members of the British royal family, Prime Minister Indira Gandhi of India, and many other foreign dignitaries. The Champ de Mars is also a site of national significance. Mauritius independence was proclaimed here in 1968.

Another important site in Port Louis with national significance is the Jardin de la Compagnie, or company garden of the French East India Company. Located in the center of downtown this public garden was originally the site of the first permanent housing built in Mauritius. It is filled with beautiful old banyan trees, fountains, and historic statues, along with numerous street vendors. While pleasant during the day, at night it attracts a sleazy crowd worth avoiding.

Directly across the street from the Company Garden is the historic Mauritius Institute building, home to the Natural History Museum. I was eager to see this museum that sounded like the best if not the only place in

Mauritius to learn the story behind the dodo. The museum has one of the most extensive collections of dodo bones anywhere in the world including the only reconstructed dodo skeleton made from bones of a single bird.

I expected to see an impressive display of exhibits about the dodo and other extinct creatures befitting the importance of the dodo's association with the very identity of Mauritius. Instead, I found only sad scraps and tired bones of the real bird. There was nothing close to the perky national icon that can be seen everywhere in Mauritius.

CHAPTER 4

HOW DEAD IS THE DODO?

MY HOPES WERE HIGH for the Natural History Museum in Port Louis. The natural history of Mauritius is amazing, exotic, and unique. This island was home to the dodo, the most famous extinct species of modern times. After dinosaurs, the dodo is probably the most famous extinct species in history. In addition, there were colossal lizards, tortoises, and parrots roaming the isolated Mascarene Islands.

The Mauritius Institute building was quite impressive on the outside. Gothic lettering proclaiming that this solid yellow structure is indeed the Mauritius Institute is clearly visible through black iron gates from Le Chaussee

14. 1606 drawing of a dodo by Carolus Clusius.

Street. The building, with white columns and arches, was modeled after the Colombo Museum in Sri Lanka and opened its doors in 1885. The architecture is representative of the grand colonial tropical style that was seen throughout the British Empire during its zenith in the nineteenth century.

Alas, the inside of the museum did not equal the grand shell. The exhibits seemed tired and mothballed. Most of the contents also dated from the heyday of the British Empire and looked like it. Large marine specimens such as squids and eels were decomposing in glass tanks of formaldehyde and some of the taxidermy sea turtles and dolphins had seen better days.

Three main galleries present the natural history of Mauritius over the past 500 years. The first gallery contains exhibits of fishes and shells. Hanging from the ceiling are large taxidermy specimens of several giant sea bass and sharks, all of which are starting to disintegrate. Against the wall are large glass display cases filled with assorted fish and shell specimens.

The center of the second gallery features large taxidermy specimens of Leathery Turtles a.k.a. *La Tortue Luth* (Dermocheys coriacea). They look very leathery indeed and seem to have lost much resemblance to living turtles. One wall is devoted to corals, minerals, and fossils displayed in large glass cases. On the opposite wall are displays of mounted butterflies and insects along with a large aquarium filled with very dark water that has no signs identifying what it contains. The contents proved to be a few small, very ordinary fish, and oddly, a miniature stone Easter Island head. The aquarium looks like it belongs in someone's living room instead of a museum.

The third gallery is dedicated to the World of the Dodo and I expected this to be the crown jewel of the collection. I suppose it was in fact the highlight of the museum but disappointing nonetheless. It's a large room that seems only partially filled, with a chipped tile floor in need of cleaning, and the air conditioning is provided by a lonely antiquated electric floor fan wheezing away in one corner.

The center of the room has a giant table that holds three small glass display cases, each containing a complete dodo skeleton. To one side of these is a long flat display case containing what appeared to be about two dozen dodo thigh bones and about eight pelvis bones. There is no sign identifying or explaining these so I can't be sure. Also on the large table is a skeleton of Poule Rouge, the Red Hen, another extinct flightless bird unique to Mauritius.

Next to the entrance to this gallery is a large glass display case containing a large, perhaps life-size reproduction of a dodo covered in feathers that look suspiciously similar to chicken feathers. In fact, I'm pretty sure they are chicken feathers. This very old model of the dodo with an oddly symmetrical oval body is placed next to a large oil painting of a dodo.

The single concession to modern technology in the museum is a video monitor in the dodo gallery playing a continuous videotape, narrated in both Dutch and English, that documents an excavation of dodo bones by the 2009 Dodo Research Programme.

On the north wall is a glass case containing various taxidermy specimens, including examples of introduced predators that decimated the dodo population. The case contained a monkey and three rats. Finally, there is a glass display case containing postmarked envelopes with Mauritius postage stamps that included a dodo as part of the design.

15. *"The Hooded Dodo,"* 1803

My impression after repeated visits is that for some reason this museum has been neglected, underfunded, or perhaps both, for a long time. It's a shame because it has the potential to become a first class museum that could attract large numbers of visitors and be a focal point for ecotourism in Mauritius. I wonder what treasures might be found in the large collection of artifacts the museum has in storage that have never been displayed.

The dodo exhibits were nonetheless fascinating and cemented my interest in this odd bird. The Natural History Museum exhibits helped me understand the scientific story behind the dodo and how it came to evolve on Mauritius in the first place. The reassembled dodo skeletons, bones, paintings, and even postage stamps gave me a better idea of what real dodos looked like when Dutch explorers arrived on Mauritius in 1598.

Based on descriptions I have read, I imagine that these first European explorers would have found an island thickly covered with ebony forests filled with thousands of dodos the size of large turkeys. The dodos had no fear of humans because they evolved in the total absence of predators, and would not have run from very hungry sailors who could easily approach them and club them to death. Unfortunately for the dodos, this is exactly what happened. A similar fate met the giant tortoises the explorers found wandering about the island. Giant tortoises were loaded onto ships for long sea voyages and would stay alive as a source of fresh meat until the day they were needed for the crew's menu du jour.

Our knowledge of what the dodo looked like comes largely from written records made by eyewitnesses and paintings of live birds that were exported to Europe from Mauritius before 1630. These provided incomplete and often inconsistent descriptions of what the dodo looked like. We don't know how it lived, what it ate, or other basic information. Ironically, there are more complete fossil records available for dinosaurs that lived millions of years ago than there are for the dodo. The physical remains of the dodo available to researchers today consist only of assorted bones, tissue fragments, and a few complete skeletons.

There is considerable disagreement as to what the dodo actually looked like. European paintings of living specimens made in the early eighteenth century depict the dodo as a very fat, ungainly bird, an image that endures in the popular imagination. Recent studies suggest that the dodo was in fact much slimmer than previously thought and wasn't the "plus size" bird usually depicted in books and illustrations.

The European paintings before 1630 may have been accurate images of dodos that made it to Europe. But they did not represent what wild dodos looked like in Mauritius. The few live birds that survived the long voyage to Europe may have become fat from overfeeding and confinement.

16. Ostriches and a dodo, 1805

The first Dutch expedition to Mauritius in 1598 produced drawings of a dodo-like bird that show a thin, much trimmer bird than those portrayed in later paintings. Recent studies of dodo bones suggest the dodo was not as heavy as previously imagined.

The Oxford Natural History Museum and Cambridge Zoology Museum measured dodo bones to calculate how much weight a living bird could carry, and the results determined that the fat dodo portrayed in later paintings could not have been supported by its skeleton. This supports the hypothesis of a slimmer bird.

In additions to paintings and sketches, there are written descriptions recorded by Europeans who saw living specimens brought to Europe. One of the most famous of these was recorded by the English theologian and historian Sir Hamon L'Estrange who observed a live dodo in London in 1638:

"About 1638, as I walked London streets, I saw the picture of a strange looking fowle hung upon a clothe and myself with one or two more in company went in to see it. It was kept in a chamber, and was a great fowle somewhat bigger than the largest Turky cock, and so legged and footed, but stouter and thicker and of a more erect shape, Coloured before like the breast of a young cock fesan, and on the back of a dunn or dearc colour. The keeper called it a Dodo, and in the ende of a chimney in the chamber there lay a heape of large pebble stones, whereof hee gave it many in our sight, some as big as nutmegs, and the keeper told us she eats them (conducing to digestion), and though I remember not how far the keeper was questioned therein, yet I am confident that afterwards she cast them all again."

Based on all the remaining paintings, sketches, contemporary written records, and physical bone fragments, we have an approximate idea what a dodo looked like when the Dutch first arrived on Mauritius in 1598.

The dodo was a large bird, about three feet high, and weighed about 50 lbs. It had a very large, slightly hooked beak and head with a mostly unfeathered face with a short neck. Its body feathers were possibly grayish brown in color and its tiny flightless wings were possibly yellow. It had short curly tail feathers, perhaps gray or white in color, at the thickest part of the rump. We don't know what the dodo ate but its huge beak was well suited for crushing fruit. Its diet most likely included fruits, seeds, and berries.

I learned that the dodo was just one of many unusual species unique to Mauritius that have become extinct. These include a number of flightless birds, parrots, giant tortoises, a giant skink, and other lizards. Skeletons and pictures of some of these extinct species are on display at the Natural History Museum in Port Louis.

Each of the three Mascarene Islands had its own version of a large flightless bird like the dodo though the true dodo was found only on Mauritius and as such called the Mauritius Dodo. The Rodrigues Solitaire

(*Pezohaps solitaria*) and the so-called "White Dodo of Réunion" (*Didus borbonicus*) are commonly known as the other Mascarene Dodos, first identified as such by the Portuguese and other early visitors to these islands.

I was somewhat shocked to learn that the dodo was a member of the pigeon family and, in effect, a giant flightless pigeon. The very idea conjured up alarming visions of me wading through crowded flocks of annoying pigeons in the busy streets of American cities as I have done on countless occasions, only these birds would be the size of turkeys with beaks to match.

In the 1840s, Theodore Reinhardt at the Royal Museum of Copenhagen was the first to make the suggestion that the dodo was a member of the pigeon family. In 1848, Hugh E. Strickland and A. G. Melville affirmed the relation between the dodo and pigeon ancestors in their book, *The Dodo and Its Kindred.* Strickland was the president of the Ashmolean Society in Oxford, whose collection included the famous surviving examples of a dodo's head and foot. The two authors were interested in the dodo and sought to understand the reasons for its extinction.

The two scientists conjured up a decent theory. Current DNA tests performed on surviving samples of physical remains in 2002 have shown that the closest living relative to the Mauritius Dodo and the Rodrigues Solitaire is the Nicobar pigeon, which comes from Southeast Asia. The Nicobar pigeon still exists today and is known by the scientific name of *Caloenas nicobarica*. The dodo and the Solitaire shared a common ancestor that separated from the Nicobar pigeon more than 40 million years ago when it began island-hopping across the Indian Ocean.

The Mascarenes were never connected to land and the only way for animals to reach these remote, isolated volcanic islands was by flying, swimming, or hitching a ride on some kind of floating debris like tree trunks. Birds and bats reached the islands by drifting winds. They brought seeds with them, which helped to colonize plant life. Once life became sufficiently established, species could make the short hop to nearby islands.

This helps to explain the existence of related but different endemic flightless birds on Mauritius and Rodrigues Island. The Mauritius Dodo and the Rodrigues Solitaire both belong to the same family Raphidae, descended from an ancient Nicobar pigeon, but once separated on different islands, they evolved into two distinct birds.

The only mammals endemic to the Mascarene Islands are fruit bats that were able to fly there under their own power, if inadvertently. Presumably, bats reached Mauritius millions of years ago when an early ancestor flew off course during a storm or hitched a ride on a clump of floating debris.

No one knows how or when the dodo reached Mauritius, which is thought to have emerged from the

17. Engraving of an ostrich, dodo, and cassowary dated 1822

sea as a result of volcanic activity around eight million years ago. The dodo could have reached Mauritius by flying, swimming, or by riding floating debris from other islands. Once on Mauritius, the pigeon ancestor of the dodo found itself on an island with no predators, plenty of food, and gradually began to adapt to its new environment. Over time, it lost the need for flight, developed short stubby wings, a giant body, and a huge, hooked beak.

The dodo's evolution from a flying pigeon into a large, flightless, and totally strange-looking bird over millions of years is one of the most famous examples of the biological phenomenon known as island gigantism. Gigantism occurs on isolated islands where animals grow larger due to a lack of predators and competition than would be present on mainland environments. The two species of giant tortoise that lived on Mauritius with the dodo are other examples of gigantism. The complementary phenomenon to island gigantism is island dwarfism where large animals become reduced in size in response to constraints imposed by living in a very small environment. Island dwarfism has occurred with dinosaurs and mammals, notably elephants.

I hadn't heard of this theory until I read about it in connection with the dodo after my first trip to Mauritius. It certainly seems plausible to me and the dodo is a graphic if not cartoon-like example of island gigantism supported by DNA samples and other published scientific evidence.

Islands serve as natural laboratories where mutations can evolve and flourish more dramatically than they might on continents. Mauritius and the other two Mascarene islands were discovered in what scientists call pristine condition. That is to say, they were uninhabited by humans and had no history of human occupation. This is an important distinction between other isolated islands that were discovered by Europeans late in recorded history like Madagascar and New Zealand, which were occupied by aboriginal people.

The natural laboratories on the Mascarene Islands fostered unique creatures without man interfering for thousands of years. The transformation of the ancestral flying Nicobar pigeon into the large, flightless and decidedly strange-looking dodo is one of history's more visually striking examples of pristine island evolution. But the dodo wasn't the only strange bird.

Early visitors to the other Mascarene islands noticed that each had their own endemic flightless bird similar to the Mauritius dodo. The Solitaire of Rodrigues Island had a very different physical appearance from the Mauritius dodo but was in fact closely related. Both birds were scientifically classified as members of the same subfamily Raphinae. For many years, large numbers of people believed there was a "White Dodo" that lived on Réunion. It was also flightless, and similar in appearance albeit with white feathers.

Travelers' accounts of large, white-colored flightless birds on Réunion generated much literature about these mythical birds. Several paintings of white dodos painted in Holland during the seventeenth century further fueled speculation. Yet, evidence supporting the existence of a White Dodo is much more sketchy than descriptions and illustrations that are available for the Mauritius Dodo and the Rodrigues Solitaire.

Live Mauritius Dodos were sent to Europe in the early seventeenth century where they were extensively illustrated and written about. The Rodrigues Solitaire was described in detail by the noted explorer and author François Leguat in his writings. During his extended stay on Rodrigues, he made a drawing of a live bird, the accuracy of which is supported by recent scientific analysis of Solitaire bones. Documented descriptions of the "Réunion Dodo" were much more vague and elusive. Analysis of fossil bone discoveries in the 1970s revealed that the "White Dodo" was probably an ibis (*Threskiornis solitarius*) and not related to the Mauritius dodo.

Sadly, all three "Dodos" became extinct after the arrival of humans.

The possible existence or nonexistence of a White Dodo notwithstanding, I became much more interested in Réunion during the course of conversations I had with several Mauritians. I was even asked why I wanted to visit Réunion at all, since from this gentleman's point of view, compared to Mauritius it is very expensive, the beaches aren't as nice, and it is after all, French. Oui, I wanted to see it even more after hearing that.

CHAPTER 5

LA RÉUNION

DURING THE 45-MINUTE FLIGHT from Mauritius to La Réunion's main airport near Saint-Denis, I sat next to a convivial Air France pilot. He was going to fly his 747 cargo plane back to Paris as part of his regular run after spending several days of R and R in Mauritius. He told me that I would love La Réunion and that the people there are very friendly and welcoming. He was right. The island, which is the largest of the Mascarene Islands, was spectacularly scenic with the highest mountains and the only active volcano in the western Indian Ocean. The Réunionnais, the people of Reunion Island, are indeed very friendly. English-speaking, non, they are not.

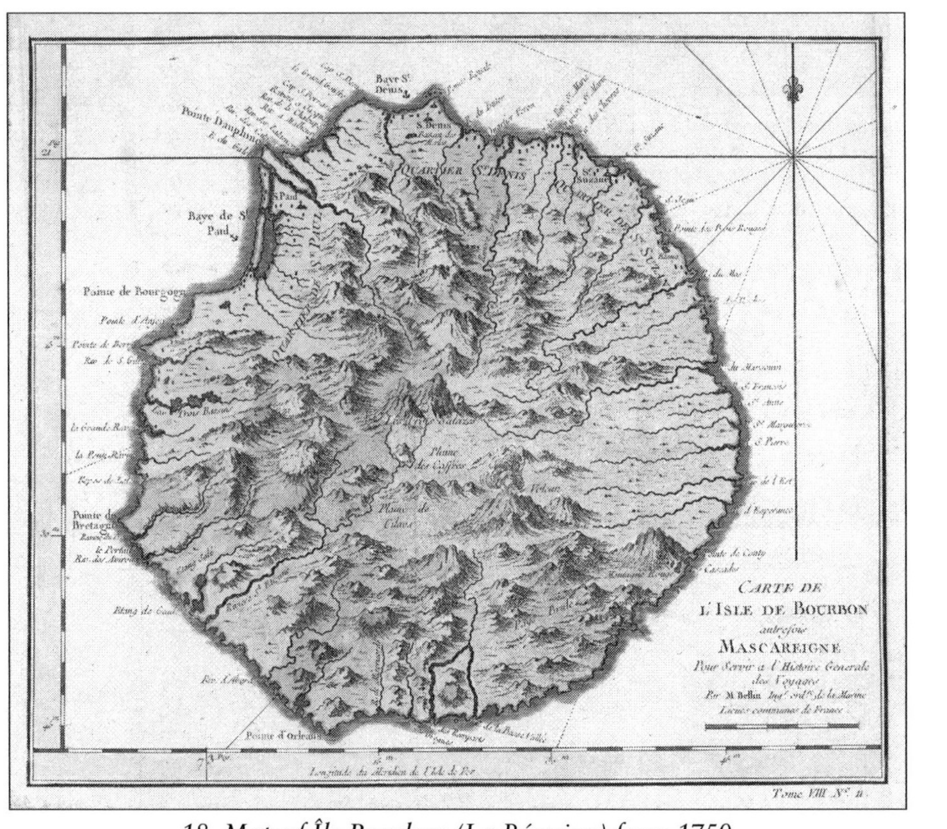

18. Map of Île Bourbon (La Réunion) from 1750

It didn't hurt that I was American, a tourist species rarely seen on La Réunion. The Mascarenes are definitely not on the radar as a vacation spot for most Americans. And La Réunion is often referred to as a travel secret that the French keep to themselves. It's certainly not a secret in France where popular direct flights from Paris lure huge numbers of tourists to La Réunion for vacation year round, many of them deciding to retire there. The population of the island is around 800,000 and growing.

Politically, La Réunion is a Départment Français d'Outre Mer, equivalent to a départment in mainland France with representation in the French Parliament, similar to the political status that the U.S. state of Hawaii shares with mainland American states. There are only three other French Départments d'Outre Mer: Martinique and Guadelupe in the Caribbean, and French Guiana north of Brazil in South America. La Réunion is by far the most geographically remote of the four and is in fact the furthest outpost of the European Union.

Unlike Mauritius, La Réunion has neither large resort hotels, nor any five star hotels at all, and it lacks the stunning white sand beaches. Hotels are generally small with the largest of these in the cities of St. Denis and St. Pierre. Accommodation options elsewhere are usually modest and include gîtes or small inns, chambres d'hôtes, apartment rentals, and camping. La Réunion does have a busy resort nightclub scene, mainly in the western towns of St. Gilles Les Bains and L'Hermitage Les Bains. I think the majority of visitors to La Réunion come to see the stunning natural scenes, especially the beautiful, rugged mountains that occupy the interior of the island.

Such a setting is a natural for extreme sports. Adventure sports are very popular on La Réunion and include hang-gliding, paragliding, skydiving, white-water rafting, kayaking, and mountain biking. There are well-marked sentiers or footpaths throughout the island that offer jaw-dropping views of soaring mountains, waterfalls, subtropical forests and some of the most perilous roads I have ever driven. The volcano, Piton de la Fournaise (Peak of the Furnace in English), is frequently active and is reasonably accessible by car or bus. Unlike the other two Mascarene Islands of Mauritius and Rodrigues, La Réunion retains much of its native forest. More than 100,000 hectares or 40% of the island lies within the protected borders of the La Réunion National Park. This provides an opportunity to see and imagine what the Mascarene Islands looked like before humans arrived.

St. Denis is the largest city and center of government. Unlike its Mauritian counterpart of Port Louis, it has no towering skyscrapers, only the giant minaret of the local mosque. This city of around 140,000 looks like a French provincial town and strikes me as a combination of Paris and Hawaii, two of my favorite places. There are many small shops, restaurants, and brasseries just like those found in mainland France, but here there are also beautiful tropical flowers and palm trees everywhere. St. Denis is filled with beautiful historic Creole architecture, small streets crammed with French cars, and lots of beautiful aging and sometimes decaying mansions hidden behind large iron gates covered with bougainvillea. Everything is priced in Euros and is very expensive, especially compared to Mauritius.

The French manage to keep La Réunion a secret in large part because virtually no one speaks English and there doesn't seem to be much tourism outreach to attract Anglophone visitors. Surprisingly few people in La Réunion speak English and I think it would be a frustrating place to visit for anyone who doesn't speak at least a little French. La Réunion was a pleasant surprise for me but I definitely want to recharge my French language skills before I return again.

However, the language hurdle did not stop me from enjoying the French ambience or food. My first dining experience in St. Denis was at a brasserie that looked exactly like a Parisian bistro complete with white tablecloths, stone walls, and formally clad servers. I could have been in Paris except for the stunning tropical flowers on my table, the welcoming Malagasy waitress, and fantastic dishes prepared with exotic locally grown epices (spices). I later discovered wonderful boulangeries (bakeries), ice cream parlors, and small bistros with menus written on sidewalk chalkboards, and other culinary pleasures exactly like those found in France.

19. Monument aux Morts and Hôtel de Ville (City Hall), Saint-Denis

Like Mauritius, La Réunion has many vehicles for a small island which create major traffic jams but the sight is quite different. On Mauritius it's very common to see antiquated trucks, cars, and especially motorcycles belching clouds of thick black exhaust smoke. On La Réunion, the vast majority of cars and trucks are very modern, usually French, and comply with EU pollution controls. Traffic can be very intense at rush hour, and crawls through the streets of St. Denis but without the feeling of chaos that accompanies rush hour traffic in Port Louis.

La Réunion's multicultural mix is different from that in Mauritius. People of African descent make up nearly half the population of La Réunion, and Hindus of Indian descent make up about a quarter whereas they are the majority on Mauritius. Muslims make up a small minority of the overall population but have a visible presence in many communities. The majority of Réunionnais are Roman Catholic.

French culture is omnipresent in La Réunion, evident from the architecture, the food, and even the stylish clothes people wear. On La Réunion, Muslims wear traditional garments but it often seems like everyone else just stepped off a plane from Paris and many of them have. Americans might be astonished to find this little piece of France planted on the other side of the world in a remote part of the Indian Ocean. Even more surprising is the fact that La Réunion has been an integral part of France for more than 350 years.

The Portuguese are believed to have been the first Europeans to discover La Réunion in 1507 and made subsequent visits but never settled on the island. France first claimed the island in 1642 and in 1649 named it Île Bourbon after the French royal family. Over the years, it has been known by several different names. After the French Revolution, it was renamed La Réunion. Aside from a few years after the temporary conquest by the British in 1810, La Réunion has been continuously occupied and ruled by France since 1665.

The style of architecture of the island is unmistakably French whereas the style of Mauritian architecture reflects its multicultural Dutch, French, and British past. St. Denis was founded in 1668 and is filled with historic architecture, from grand public buildings and monuments to quaint Creole colonial houses. I spent many days just walking through the streets and parks of St. Denis. It is a very accessible by foot.

I followed a route from Le Barachois, the public park along the waterfront, to the Avenue de la Victoire, which turns into Rue de Paris, which in turns ends at the Jardin de l'Etat. One of the more beautiful buildings along this route is the old city hall or Hôtel de Ville, an impressive nineteenth century building painted bright yellow and decked with French flags and plaques marked RF for Republique Francaise. This lovely building contrasts

20. Jardin de l'Etat (Garden of the State), Saint-Denis

sharply with the new city hall a few blocks away, an uninspiring 1960s concrete office block. Opposite the old Hôtel de Ville is the Monument aux Morts, a towering spire built in 1923 to commemorate local casualties of World War I. One famous WW1 veteran is Roland Garros, the namesake of the St. Denis international airport. Garros was a native of St. Denis and an aviator famous for making the first nonstop flight across the Mediterranean Sea.

I stopped at a small museum along the Rue de Paris, the "Musée Léon-Dierx" that houses La Réunion's most important art works. The museum building is quite elegant albeit small, with only enough exhibition space to display a portion of the permanent collection at one time. The museum exhibits historical and contemporary works by local Réunionnais and French artists, and by others including Pablo Picasso. There is one section dedicated to historical paintings, and includes a wall of paintings of early Reunion Island settlers, men and women painted in the nineteenth-century Whistler's Mother school of portraiture with dour expressions. The most interesting paintings to me from this section were the landscapes of Reunion Island from the mid-nineteenth century.

Especially notable is "La Cathedrâle de Saint-Denis," 1877, by Antoine Louis Roussin (1819–1894). "La Cathedrâle de Saint-Denis" is fascinating because it depicts the same cathedral that is still standing today, just down the street from the Leon Dierx museum along the Avenue de Paris. You can compare the scene from 1877 to the same scene today. The 1877 painting shows detailed images of people including women in saris, European women wearing Victorian dresses with bustles, and men in top hats.

At the end of the Rue is the Jardin de l'Etat, a large public garden that houses the Muséum d'Histoire Naturelle or Natural History Museum. The Natural History Museum has a large exhibit of wildlife specimens peculiar to La Réunion that I found quite interesting despite the lack of English language signs or English-speaking staff. I was especially interested in a display of an egg from the extinct elephant bird (*Aepyornis maximus)*, an extinct flightless bird that stood 3 meters high (10 feet). The egg was displayed next to an ostrich egg that looked like a chicken egg by comparison. In fact, a single elephant bird egg contained the equivalent of 140 chicken eggs. That was a seriously big bird.

The elephant birds were native to Madagascar and are believed to be the largest birds that ever lived. Sadly, like the dodo, they survived just until the seventeenth century even though they were frequently sighted by early French officials living in Madagascar. It was a fascinating yet troubling reminder of the extinction of yet another exotic bird species in the Indian Ocean region.

One curious and noticeable aspect of La Réunion is that the most popular local beer is called Bière Bourbon and uses the picture of a smiling Mauritius Dodo as its symbol. There are large advertising signs for this beer all over the island, known locally as Dodo beer. Ask for a "Dodo" at a bar and this is what you will get. This seemed odd to me since the dodo was not from La Réunion and there is no local beer on Mauritius called "Dodo" where the dodo is so famous. Purely for the purposes of scientific research, I discovered that Dodo beer is in fact quite good. The same goes for Phoenix beer, the most popular brand brewed in Mauritius.

La Réunion seems much more prosperous than Mauritius, thanks to heavy subsidies from the French government. Roads and public infrastructure are modern

21. Hôtel de Ville (City Hall), Saint-Pierre

and built to European standards, including the spectacular Routes des Tamarins linking the cities of the west coast of the island with soaring bridges and deep tunnels. Opened in 2009, this highway was one of the most expensive roads ever built in the European Union.

Public transportation on La Réunion is very good and includes an island-wide bus system called Cars Jaune that connects most towns using sleek bright yellow buses. A shuttle service called the Navette links the Roland Garros International Airport with St. Denis using equally modern vehicles. None of the current public works projects and European-standard level of services found on La Réunion would be possible without significant financial support from the French government.

After exploring St. Denis for a few days, I headed down the west coast to Saint-Pierre, stopping overnight in St. Leu and exploring a few sites along the way. I always have my eye out for something unusual and La Réunion did not disappoint.

I have to admit that one of the highlights of my visit to La Réunion was discovering the Garden of Eden. I have always wanted to see the Garden of Eden, I knew it existed somewhere and while I had to travel to the other side of the world to find it, here it was! Imagine my surprise upon seeing a large road sign that said Jardin d'Eden south of the town of L'Hermitage. Despite its overreaching name, the small privately owned Jardin is in fact a genuine botanic garden with hundreds of different species of plants on display along with some native chameleons. No Adam, no Eve, no snake though. And no apple.

The Jardin d'Eden paled in comparison in terms of scale and setting to the Conservatoire Botanique Nationale des Mascarins located further south near the seaside town of St. Leu. This impressive botanical garden is set in the hills and has acres filled with both native and regional plants, all identified with labels. The Conservatoire is set on the ground of a nineteenth century French colonial mansion and has a spectacular view overlooking the Indian Ocean. For those interested in botany, it's well worth a visit.

St. Leu is a mecca for paragliding due to favorable uplifting thermal winds that allow gliders to float for ages before eventually landing near the beaches far below. There are a number of paragliding operators in St. Leu that use a launch site in the hills above the Conservatoire. Shortly after launching, many of these fly directly over the Conservatoire at low altitude, which is a startling sight for the unprepared. While alone in a garden trying to read labels of plants in complete solitude, I heard strange voices from above.

Just for the record, I am not given to hearing imaginary voices. I looked up and saw six or seven brilliantly colored parasails only 10 meters above me, with their occupants busily chatting away in French. Over the course of several hours, dozens of parasailers flew over the Conservatoire, filling the skies with bright parachute cloth.

Saint-Pierre is the south end hub of the island and, like St. Denis, is home to many beautiful historic buildings. Despite having a population of only about 30,000, it has a very busy nightlife scene, one that seemed authentic and casual. Most of the clubs, restaurants, and bars are concentrated on Boulevard Hubert Delisle that borders the seafront. A pleasant, albeit a bit rocky, public beach is on the other side of the boulevard. The restaurants in Saint-Pierre range from good to excellent, especially seafood establishments that serve some of the freshest fish from the cleanest seas anywhere on earth. Vanilla is grown on the island and I tried several fish dishes prepared with fresh vanilla, which were quite good, especially the swordfish.

After every meal I was offered a *rhum arrangé*, a Réunionnais specialty consisting of local rum infused with spices or fruit.

I enjoyed exploring both St. Denis and St. Pierre, two cities I had not expected to be so interesting and filled with such intriguing historic architecture and great food. However, most people come to La Réunion to experience its natural wonders, particularly the Les Cirques and the Piton de la Fournaise volcano. These are located in the sparsely populated mountainous interior of the island and are fairly inaccessible from the coastal cities and towns. I was to discover just how inaccessible even when utilizing modern transportation in the form of a zippy Renault Megane rental car.

CHAPTER 6

LES CIRQUES

THE FRENCH SETTLERS OF LA RÉUNION certainly gave descriptive monikers to the geological wonders of their island. One of the world's most active volcanoes, Piton de la Fournaise translates to the Peak of the Furnace. Piton des Neiges in English is Peak of Snow. Not a bad title for the dormant volcano that rises almost two miles from sea level and is the highest point this side of the Indian Ocean. Some poetic license was taken with the christening of Piton des Neiges, however, since snow rarely falls so close to the equator.

Then there are the three giant cliff-rimmed natural amphitheaters called cirques, which means circus in French. They are in fact a geological three-ring circus.

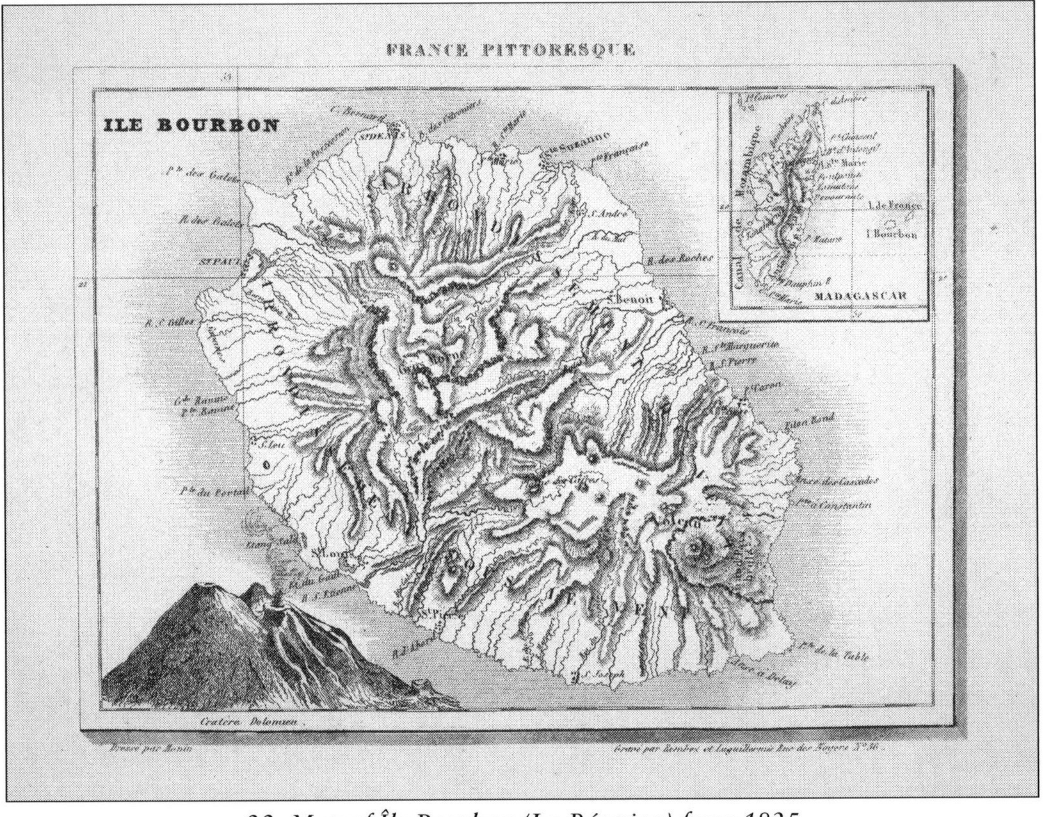

22. *Map of Île Bourbon (La Réunion) from 1835*

The cirques and the two towering volcanic peaks dominate the interior of La Réunion. Cirques are formed over millennia by erosion and heavy rainfall creating steep and dramatic gorges. The three cirques, Cirque de Mafate, Cirque de Salazie, and Cirque de Cilaos, were born eons ago when the dome of a prehistoric volcano collapsed. Les Cirques are probably the biggest single draw for visitors to La Réunion, many who spend days or weeks hiking and exploring the unique landscape.

Hiking is more of a past tense activity for me these days, despite the fact that I live in the Pacific Northwest, land of Mt. Rainier, numerous national parks, and endless hiking trails. There is something about a replacement knee that stops me from wanting to tackle giant volcanoes on foot. So I rented a car for some quick day trips. Now that was an adventure.

For the most part, La Réunion is an easy place to explore by rented car. The roads are excellent and the French drive on the same side of the road as Americans. I was told that it was a 37 km or 23 miles drive from the town of St. Louis, the gateway for the road to the town of Cilaos inside Cirque de Cilaos. I could easily make it there for lunch and a brief walking tour, and return in about six hours. No problem.

I wanted to see for myself what these remote forested mountain areas were like. The cirques contain deep gorges, subtropical rainforests, waterfalls, and flower-filled meadows. They are also a protected habitat for

23. Cirque de Cilaos, La Réunion

plants and animals, which have survived at a much higher rate here than on other Mascarene Islands. I think the extreme topography of the cirques helped to save them, or at least slowed down contact with humans.

The region encompassed by Les Cirques is sparsely populated and is home to only a small number of villages, the largest of which is Cilaos. I decided that Cilaos might be the easiest to visit for a day trip from St. Leu where I stayed for a few days. Cilaos is home to natural thermal hot springs and during the nineteenth century the town developed into a health spa resort for colonials seeking a cure. Today the town's economy relies heavily on tourism and still includes a hot springs spa, and a small but growing wine industry.

Local tasting rooms offer local wines made from grapes grown in nearby vineyards. I didn't have the inclination or will to sample a glass of the Vin de Pays after the harrowing drive there. The road from St. Louis to Cilaos, the N5, is described as scenic, challenging, and, according to some, conducive to nail-biting. It is definitely all of these and more. I have never had a more white-knuckled driving experience in my entire life.

This road has literally hundreds of sharp hairpin turns, including a couple of benders where the road narrows to a single lane, forcing me to drive blind along the side of a cliff. I spent virtually the entire time hoping I wouldn't collide with oncoming traffic that was invisible until the last moment. In some places the road is carved into a sheer cliff hundreds of meters above a deep ravine. I tried not to look down.

The most nerve-wracking portion of this drive, of which there were many, was driving through a couple of long, single lane, unlit tunnels that had no traffic signals of any kind to alert you of oncoming traffic. Halfway through one tunnel I saw a pair of headlights coming towards me at high speed. I frantically pounded my horn, flashed my headlights, and slowed down until the other car stopped and eventually backed up several hundred meters to exit the tunnel. I passed through to find several cars waiting to enter, followed by a tour bus that seemed too large to fit inside the tunnel.

24. Hairpin turns on the road to Cilaos

I eventually made it to Cilaos in about two hours, but was sufficiently shaken by the experience that it took a while to release my grip from the steering wheel. Driving was just an unnerving aspect of this trip. And I am not exaggerating the danger. I passed at least one serious car accident on the way, saw an ambulance extricating passengers from a wrecked car, and also noticed numerous roadside shrines to earlier accident victims at frequent intervals. There is regular bus service to Cilaos and I saw many tourists in town who must have been blissfully ignorant of the horror.

Driving issues aside, the scenery inside Cirque de Cilaos is truly stunning and lives up the picture postcards

I saw in St. Denis bookshops. The town of Cilaos is charming, and the views are spectacular, reminding me of a subtropical Switzerland. Hawaii also has some beautiful mountains but those on La Réunion seem much taller and the sweeping vistas are much more dramatic. There are many well-preserved and brightly painted nineteenth century Creole cottages and homes and gardens filled with brilliant tropical flowers.

The town is dominated by the Notre-Dame-des-Neiges cathedral completed during the 1930s, which looks out of place compared with so many older wood-framed houses dating from the nineteenth century. Walking around Cilaos, I couldn't help but get a sense of the extreme isolation early settlers must have felt living here.

The first inhabitants of the cirques were escaped African slaves, or "marrons" who took refuge in the remote cirques and formed their own communities in the eighteenth century. The cirques would have formed a perfect hiding place for them because it was virtually inaccessible. Even today, only a few roads connect the small villages of the cirques with the rest of La Réunion over deep gorges and valleys. The Cirque de Cilaos is believed to have originated with the Malagasy word Tsilaosa, which means "the place one never leaves." Before roads were built into the cirques, slaves from sugar plantations believed that they were safe in this remote high altitude refuge but sadly most were eventually tracked down and forcibly captured.

On a happier note, Cilaos is famous for its embroidery work and there is a small museum dedicated to this local tradition called Maison de la Broderie. Embroidery was introduced in the nineteenth century to give local women a productive craft to occupy their time during many long days in the isolated confines of the cirques. It developed over time into a distinctive regional style of embroidery.

After lunch at an inn, I took in the local sites by foot, and decided that I didn't want to drive back to the coast on the same road. Big problem. I discovered that there is no other road out of Cilaos to St. Louis. I made it safely back thanks to a couple of hours of intense driving concentration but it took me a while to recover. Many local inhabitants drive this route on a daily basis and probably don't give it a second thought, a feat that boggles my mind. The cirques are indeed lovely, well worth visiting, but should I come back, I don't want to drive myself.

Yet, I really wanted to see the Piton de la Fournaise even if it meant driving over dangerous roads. For many people, a visit to the Piton de la Fournaise volcano is the single most important reason to visit La Réunion. This is one of the most active volcanoes in the world and attracts thousands of visitors during periods of eruption. These can last several weeks at a time during which access to the volcano is restricted.

I learned that serendipity or lack thereof determines the success of a visit to Piton de la Fournaise. The volcano might be erupting and you might be able to see it from the edge of the caldera—or not. At 2,631 meters (two miles plus) above sea level, I was hoping to see it up close or at least from a reasonable distance. I got up my courage after my "cirque of death trip" to hop back into the car and take in the view. Unfortunately, I chose to visit the volcano on a day it was shrouded by thick fog. Despite this, my drive from St. Pierre to the visitor center at Pas de Bellecombe, located at the rim of the volcano's caldera, was worth the effort. I passed through a dramatic variety of wildly different landscapes in less than two hours, from sunny beaches to subtropical Alpine forests to barren lunar deserts of lava.

Like many visitors to the volcano, I first headed to the Maison du Volcan visitor center at Bourg Murat located in La Pleine des Cafres, a somewhat desolate area about 2,000 meters or 1.25 miles above sea level. From there it's a 45-minute drive to Pas de Bellecombe where there are paths that lead to the edge of the volcano.

The N3 road from St. Pierre to Bourg Murat passes directly through the curiously named town of Le Tampon, and through several other towns named for their distance from the sea such as Le Quatorizième (The Fourteenth i.e. 14 km from the sea) and Le Dix-Neuvième (The Nineteenth i.e. 19 km from the sea). I stopped at the

Maison du Volcan, a combination museum and visitor center that has a wealth of information about the volcano. I decided to skip the complimentary documentary film about the volcano that was only in French, free of pesky English subtitles.

The drive from the Maison du Volcan to Pas de Bellecombe passes through small villages in lush mountain valleys that are home to small dairy farms filled with fat, happy cows wearing cowbells. I felt as though I were in a scene from the story of *Heidi*. It was amazing to think that instead of being in the Alps, those cows were on a volcanic island in a remote part of the Indian Ocean east of Madagascar. The road to the volcano descends into the La Pleine de Sables which looks and feels like a lunar landscape, or perhaps Martian since the sands are reddish in color. Here there are no trees or life of any kind and no paved roads. You drive your car directly on lava all the way to Pas de Bellecombe, which made for an otherworldly experience. The unpaved hard lava roads were surprisingly easy to drive on which was a welcome earthly surprise.

I arrived at the Pas de Bellecombe visitor center to find the entire area covered with thick fog. It was impossible to see the volcano and rather than wait many hours for it to clear, I headed back toward St. Pierre. Not seeing the Peak of the Furnace this time around just gave me another excuse to revisit La Réunion in all its French glory.

CHAPTER 7

POSTAGE STAMPS AND PROSE

WHILE THE DODO IS JUST ONE of a number of odd natural anomalies, there are also some of the manmade variety of eccentricities calling Mauritius home. One of the more famous oddities is the Blue Penny stamp, one of the world's most valuable stamps prized by collectors and worth millions.

The Blue Penny stamps even have their own museum and it is definitely the best museum in Port Louis. And I'm not saying that just because it has air-conditioning and a great bookshop.

In fact, I nearly missed the Blue Penny altogether. I walked past it several times and thought that this modest two-story building was a retail store.

25. Blue Penny Museum, Port Louis

It's tucked away in a corner of the Caudan Waterfront complex in Port Louis and easy to bypass if you aren't looking for it. Founded in 2001 as a project of the Mauritius Commercial Bank, the exhibits in the Blue Penny Museum are not limited to the history of postage stamps. They cover the entire history of Mauritius from exploration and settlement through the colonial period with sleek exhibits filled with interesting maps, photographs, and engravings. Some of the historical exhibits include sound effects, an endless loop of background noise that is supposed to reproduce the sounds of Port Louis streets in the nineteenth century.

I knew very little about the philatelic (stamp collecting) world before coming across the Blue Penny Museum but soon discovered that the namesake stamps on display represent yet another curious piece of Mauritian history. The rare Blue Penny Mauritius postage stamps, two original examples of which are displayed in the museum, represent the Holy Grail to many of the world's stamp collectors.

The first postage stamps in the world were issued in England in 1840 and revolutionized postal systems everywhere. In 1847, the British colony of Mauritius became one of the first places outside of England to issue postage stamps. The first edition of stamps included an orange-red One Penny and a blue Two Pence denominations. The engraver was supposed to print Postage Paid on the border. Instead, the stamps were engraved with the words Post Office. Two hundred or so of these stamps went out on invitations to a ball held by the Mauritian Governor's wife. Only a handful still exist today and rarely come up for sale. When a Japanese industrialist sold his collection of Mauritian stamps in 1993, it went for more than U.S. $10 million.

The story behind the Mauritius Blue Penny stamps and how they became among the world's most valuable stamps is filled with mystery and intrigue. To really dive into the subject I suggest Helen Morgan's *Blue Mauritius*. The book provides an authoritative, comprehensive, and well-written account of this story and sheds light on the rarified world of high-end stamp collecting around the world. In 1993, two of these

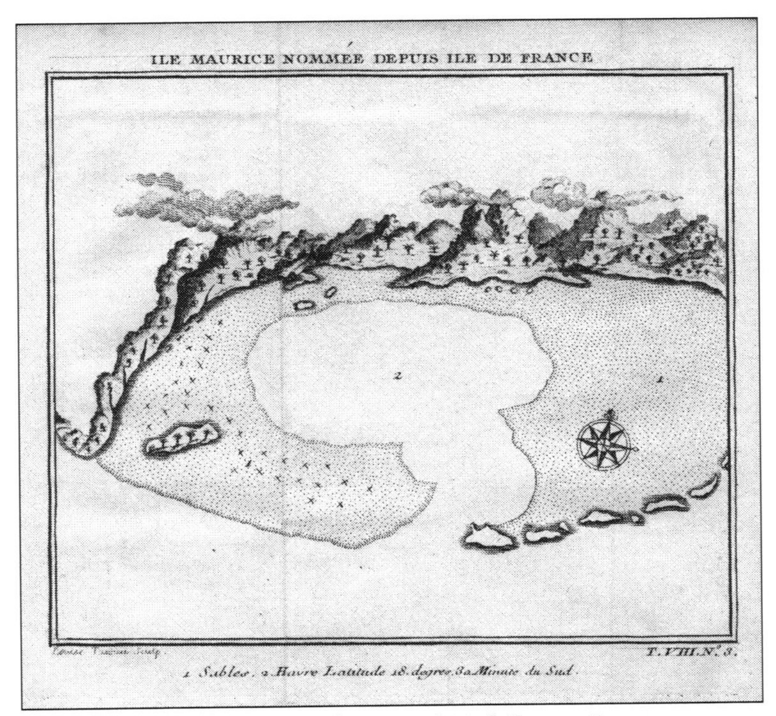

26. Early French map of Mahébourg Bay in southern Mauritius

surviving stamps became available and were purchased by a consortium of Mauritian companies for the price of U.S. $2.2 million as part of an effort to repatriate them to their native land. These stamps are now on display at the Blue Penny Museum.

Visitors take a self-guided tour through exhibits that are well documented in both French and English. Plus portable audio recorders are provided for visitors speaking other languages.

The tour starts on the second floor in a circular room that documents the human discovery of Mauritius with exhibits in chronological order. Each time period is depicted with beautiful original historic maps and a wooden model of the ships used to reach Mauritius from Europe. Mauritius has a small but thriving local industry dedicated to building scale models of wooden sailing ships for the tourist market, and there are a number of models used in several exhibits in the Blue Penny Museum to good effect.

The next series of exhibits documents the human settlement of Mauritius including the establishment and growth of Port Louis. Historic engravings of views of Port Louis are paired with modern day photographs, providing a good understanding of how much the city has changed over the years. The rest of the second floor exhibits are dedicated to the history of the postal service in Mauritius, culminating with a carefully protected display case containing the famous Blue Penny stamps along with some other valuable examples. In order to protect the stamps from the light, they are lit for only ten minutes at a time, once an hour. There are reproductions of the stamps to view when the originals are not lit.

The only permanent exhibit on the first floor of the museum is dedicated entirely to the story of Paul et Virginie in two separate rooms. The first room has exhibits and pictures of the author of the story, Bernadin de St. Pierre, and houses a large-scale model of the sailing ship St. Geran whose sinking inspired the story. There are numerous historic pictures on the walls depicting scenes from the story. In the second room is the large marble statue of Paul et Virginie sculpted by Prosper d'Epinay in 1884.

This was my first exposure to the literature of Mauritius and the novel of *Paul et Virginie* written by Bernardin de St Pierre in 1787. The story of Paul et Virginie is a tragedy about two children who grow up together, fall in love, and after being separated, want to reunite and marry. A maritime disaster prevents this from happening. Specifically, the heroine drowns because her virtue prevents her from removing her skirts and swimming to the surface when her ship goes down.

Paul et Virginie pop up everywhere, as the name of hotels, restaurants, assorted businesses, and as the object of memorials and monuments. As an English-speaking foreigner unfamiliar with historic French or Mauritian literature, I found it a bit puzzling that this anachronistic and pious story from so long ago remains popular to modern Mauritians. This is particularly so when the moralistic theme of the story seems at odds with the values and lifestyles of modern readers whose familiarity with swimming attire revolve around the bikini. I think the appeal of Paul et Virginie these days is largely sentimental.

Jacques-Henri Bernardin de Saint-Pierre (1737–1814) was a French writer who wrote two of the most famous books about Mauritius. As a young military engineer, he spent 28 months stationed on the island, which inspired his book *Voyage a l'Ile Maurice* (Journey to Mauritius) published in 1773. In this book, he describes at length the local wildlife and landscape, and also describes his horror of slavery that was finally abolished on Mauritius in 1835.

The story of Paul et Virginie is based on an actual event from 1744. The ship St. Geran was shipwrecked off Amber Island (Île d'Ambre) near northern Mauritius and two women passengers drowned because they refused to remove their clothes, the weight of which prevented them from swimming.

It's hard to imagine this story of platonic virtue appealing to audiences but one has to look at it in the context of the period during which it was written. The story made a huge impact on European readers when it was first published in 1787 and was supposedly one of Napoleon Bonaparte's favorite books. Go figure. I think a major reason that the story of Paul et Virginie remains relevant today is that it fixed Mauritius as an idyllic tropical paradise in the minds of Europeans, a notion that has endured for centuries.

Mauritius has both attracted and produced other notable writers throughout its history. Mark Twain a.k.a. Samuel Longhorn Clemens visited Mauritius in 1896 while on the last leg of a round-the-world voyage that inspired his 1897 book, *Following the Equator.* Mark

Twain is associated with Mauritius for a famous quote that you will see printed in many local advertisements and publications throughout Mauritius as well as in international guidebooks. Curiously enough, while performing scholarly research, i.e. tasting duties, I couldn't help but notice that this same quote also appears on the labels of the local Green Island Rum bottles. The quote reads:

"You gather the idea that Mauritius was made first, and then heaven; and that heaven was copied after Mauritius."

While it might be true that Mark Twain thought Mauritius was a very beautiful place, this oft-cited quote is actually part of a longer paragraph that conveys a more mixed message:

> "This is the only country in the world where the stranger is not asked 'How do you like this place?' This is indeed a large distinction. Here the citizen does the talking about the country himself; the stranger is not asked to help. You get all sorts of information. From one citizen you gather the idea that Mauritius was made first, and then heaven; and that heaven was copied after Mauritius. Another one tells you that this is an exaggeration; that the two chief villages, Port Louis and Curepipe, fall short of heavenly perfection; that nobody lives in

27. *View of Port Louis in 1844*

> Port Louis except upon compulsion, and the Curepipe is the wettest and rainiest place in the world."

Whatever his personal feelings about the place may have been, Mark Twain certainly left his mark as one of history's famous visitors to Mauritius.

The novelist Joseph Conrad came to Mauritius in 1888 as a young sea captain and fell in love with Eugénie Renouf, who declined his proposal of marriage. His experience in Mauritius inspired his novella *A Smile of Fortune*. Mauritius continues to produce noteworthy authors writing in English, French, and other languages. The winner of the 2008 Nobel Prize for Literature, Jean-Marie Gustave Le Clézio, usually known as J.M. G. Le Clézio, is from a distinguished Franco-Mauritian family and Mauritius is the setting for several of his novels. Other noted contemporary Mauritian writers include Barlen Pyamootoo and Nathacha Appanah.

One of the surprising things about Mauritius is how the French language dominates modern culture and media even though Great Britain ruled the country for the last 168 years before independence in 1968. This was a big disconnect for me and I thought it might be due to some accident of history. It turns out that that the dominance of the French language was not an accident but a result of a deliberate series of decisions by the

28. Aapravasi Ghat, Port Louis

first British governor whose actions were instrumental in preserving French culture.

British rule began in 1810 although the formal terms of control were not settled until 1814 when the Treaty of Paris was ratified. This treaty gave Île de France and Rodrigues Island to the British and returned La Réunion to the French. The British changed the name of the island back to Mauritius, established their own government, and English became the official language.

Robert Farquhar was appointed the first British Governor of Mauritius in 1810. His generous terms of French capitulation and independence from the government in London won over the French settlers on the island. French inhabitants who did not want to stay under British rule were allowed to return to France with all of their possessions. Those who wished to remain were allowed to keep their language, laws, local customs, religion, and property. This placated the French residents and kept the French language and culture alive to this day.

Governor Farquhar also encouraged planters to grow sugarcane instead of coffee, cotton, and other crops since sugarcane can survive cyclones. He was instrumental in transforming Mauritius into an economically prosperous British colony. He wasn't as successful in enforcing the abolition of slavery on the island which continued for many years after it had been outlawed throughout the British Empire by Parliament in 1807.

British efforts to abolish slavery on Mauritius faced intense resistance from sugar plantation owners. Franco-Mauritian farmers needed more labor for their expanding sugar plantations and considered slaves to be their personal property. The planters demanded compensation to be paid if the British abolished slavery.

After years of negotiations, the British Government reached a settlement with the island's slave owners, paying them two million pounds in 1833 and by 1835 all slavery was abolished on Mauritius. The terms of the law abolishing slavery forced the newly freed slaves to continue working for their former owners for a period of four years. Not surprisingly, after this period former slaves left the sugar plantations, creating a serious labor shortage.

A system of indentured labor replaced slavery as workers were imported to Mauritius from other countries, mostly from the Indian subcontinent, as well as Madagascar, Mozambique, and China. The system of indentured servitude was slavery all but in name and labor conditions were appalling. The indenture system in Mauritius originated in 1834 and was not formally abolished until 1924.

The abolition of slavery under the British brought large numbers of workers from all areas of the Indian subcontinent to Mauritius to work on sugar plantations. From 1849 to 1923, more than 450,000 came to the island under the indenture system. Indo-Mauritians gradually became the largest single ethnic group and today nearly 70 percent of the population has Indian ancestry.

Many of these arrived in Mauritius at the immigration depot on the Port Louis waterfront where they were processed by British authorities before being sent to work on sugar plantations. While immigrants also came from eastern Africa, Madagascar, and Southeast Asia, the vast majority arriving at the immigration portal of Aapravasi Ghat in Port Louis came from India.

This site in Port Louis has great significance for the majority of Mauritians as it marks the place where their ancestors first arrived in the country. Aapravasi

Ghat was named a UNESCO World Heritage site in 2006. In Hindi, Aapravasi means immigrant while Ghat means "the place where the water meets the land." The complex of buildings that make up Aapravasi Ghat is being restored and will eventually become the Mauritian equivalent of Ellis Island, a site where the ancestors of millions of Americans first arrived in the New World from Europe.

Evidence of the strong Indian influence is reflected in the abundance of historic sites on Mauritius with new Indian names. For example, the world famous botanic garden at Pamplemousses has been renamed the Sir Seewoosagur Ramgoolam Botanic Garden. This vast public garden is one of the most famous attractions in Mauritius and is filled with curious and fascinating stories.

CHAPTER 8

IN SEARCH OF SPICES AT PAMPLEMOUSSES

PAMPLEMOUSSES, A BOTANICAL WONDER, transforms history into fragrance and beauty. The sprawling 75-acre garden, about seven miles north of Port Louis, is one of the most intoxicating places I have ever visited. It is also one of the world's oldest botanical parks.

Beginning in 1768, the garden was used as a nursery for the study of plants introduced to Mauritius from all over the world. It is still used as a facility for serious conservation work. Today the park is also a popular place for families to spend the afternoon, strolling along broad palm-lined avenues named after prominent Mauritius residents and visitors, for example, Poivre

ILE DE FRANCE. ISLA DE FRANCIA.

Habitation des Pamplemousses, Jardin de Mr Céré.

Habitacion de las Pamplemusas; Jardin de Mr Cere

29. *Pamplemousses Garden in 1840*

Avenue, Labourdonnais Avenue, Charles Darwin Avenue, Thomas Huxley Avenue, and Shrimati Indira Gandhi Avenue. These shady walks lead to picturesque lakes, reflecting ponds, and specialty gardens. The signature plant is *Victoria amazonica*, an enormous water lily with leaves shaped like a giant green tart dish. Native to South America, the leaves of these water lilies can grow to an astounding 12 feet in diameter, creating a beautiful spectacle floating in the large formal reflecting pools.

Penetrate further into the vast greenery and you will discover a fascinating story of spices and fortune. This garden was not only key to Mauritian development but also to unlocking the spice trade in the eighteenth century. It is such an important historical site that the father of the present Mauritian nation was cremated here. Now the garden is officially known as the Sir Seewoosagur Ramgoolam Botanic Garden (SSR Botanic Garden) in honor of the first prime minister of Mauritius. Most people simply call it the Pamplemousses Garden after the eponymous adjacent village. Pamplemousse means grapefruit in French and the name of the garden may have come from a citrus plant called "Pamplemoussier" grown in the area. Pamplemoussier are thought to have been introduced to Mauritius from Java by the Dutch. Again the blend of cultures, nature, and history in Mauritius makes for a fascinating tale.

As one who never before saw or smelled spice trees growing in the tropics, my visit to the spice section of the

SSR garden was a sublime sensory revelation. I probably would have missed the spice trees altogether if I hadn't taken a guided tour led by one of many extremely multilingual guides that are on hand to help visitors from around the world. On this visit, my Mauritian guide spoke fluent French, English, German, and a bit of Italian to the group I had joined. I was the only native English speaker in the group of ten. Our guide was able to give us all a taste of what made Pamplemousses so uniquely beautiful. I have taken guided tours of the SSR garden on subsequent visits to Mauritius and always learn something new.

I enjoy the guided tours at the SSR Garden because not only do guides identify unusual plants for visitors, but they allow visitors to smell and feel the flora. Often the guide will pick up fallen leaves and crush them, allowing the exotic but familiar fragrances to waft into the air. The spice section of the garden has examples of nutmeg *(Myristica fragrans)*, clove *(Syzygium aromaticum)*, and cinnamon *(cinnamomum verum)*, all of which have been cultivated in the SSR Botanic Garden since 1770. The heavenly scents of cinnamon, camphor and allspice (*Quatre Epices*) are intoxicating. Guides will usually point out plants with unusual textures to touch.

I notice that all of the guides seem to use the same joke on their tours involving the mother-in-law plant. Every tour I have taken has been with a male guide

30. Pierre Poivre

(there are also female guides), each of whom shows his group of visitors a plant that has coarse sandpaper-like leaves that they can touch. He explains what it is and calls it the mother-in-law plant, after which he usually chuckles out loud.

Later he lets visitors touch a different plant with velvety-smooth leaves that he calls the father-in-law plant. This is followed by another burst of laughter at his own joke. I've heard this same joke from different guides and can only surmise that calling the shrub with scratchy leaves a mother-in-law plant is absolutely hilarious to local ears.

I visited the SSR Botanic Garden briefly on my first visit to Mauritius and was surprised to find on this remote and distant island such an elaborate and elegant European garden that was older than the United States. On subsequent visits and after learning more about the history of the garden, I better understood its relevance to the spice trade. At Pamplemousses a giant botanical experiment was conducted more than 200 years ago to see if there was a way to globalize spice production. The pivotal figure in this grand escapade involving Mauritius and the garden at Pamplemousses was the one and only Pierre Poivre. Yes, the name does translate to Peter Pepper in English!

But Monsieur Poivre gets short shrift in the amazing park he created. There is a Poivre Avenue in addition to a large bust of Pierre Poivre in the SSR Botanic Garden that is similar to one I saw in the Jardin de l'Etat in St. Denis on La Réunion. But I had to do some digging to find out the story behind these visual clues about Pierre Poivre. Although he was instrumental in creating the SSR Garden and played a large role in the history of Mauritius, I've never heard a guide mention his name while on tours at the SSR Garden. Not only was Poivre an expert self-taught botanist, he was also superb at learning languages, including Chinese, a skilled navigator, a capable government administrator, and a fearless traveler. He was an internationally famous author during his lifetime, and his admirers included U.S. President Thomas Jefferson. For much of his life he did all of these things having only one arm.

That's right, the man had only one arm. Early on in his life, his arm was shattered by a cannon ball during a battle at sea. But let's start at the beginning. Pierre Poivre was born in Lyon, France in 1719 and trained as a missionary in Paris. At age 20, the Paris Society of Foreign Missions, Missions Ètrangères, sent him to Canton, China where he had some run-ins with local authorities. He was jailed for a year apparently in a case of mistaken identity, having carried a letter of introduction for someone else. During his incarceration, he learned the local Chinese language well enough to impress a local Mandarin who not only freed him, but took Poivre under his patronage. He traveled throughout the interior of China and also to Cochin-China, known today as Vietnam. This gave him the opportunity to study native plants and further his interest in botany.

After two years, however, his superiors at the sponsoring mission begin to doubt his chosen religious vocation and sent him back to France. At Canton in 1745, he boarded the French ship *Dauphin* for the long trip home. Passing through unfriendly waters near the east coast of Sumatra, the *Dauphin* was attacked and captured by an English ship in the Bangka Strait. During the battle, Poivre was hit in the wrist by a cannon or musket ball, and later taken prisoner by the English. Within twenty-four hours, he developed gangrene and was saved by an English surgeon who performed an emergency amputation of his right arm under difficult circumstances with no anesthetic.

After the battle, the English deposited Poivre at the Dutch port of Batavia (modern Jakarta in Indonesia), leaving him to find his own way back to France. Batavia, the old Roman name for the Netherlands, was then the center of the lucrative Dutch spice trade that generated astronomical profits. After seizing control of the Spice Islands from the Portuguese, the Dutch ruthlessly protected their monopoly on cloves, nutmeg, and mace and didn't hesitate to execute potential smugglers.

Today we take spices for granted as a grocery store staple that is cheap and readily available. It's easy to forget that for much of human history, spices were a rare commodity more valuable than gold, available only from mysterious and far-away lands. It was the search for spices that motivated the great explorers of the fifteenth and sixteenth centuries, most notably Christopher Columbus, Vasco de Gama, and Ferdinand Magellan. These men were first and foremost spice hunters whose expeditions were underwritten by the royal families of Portugal and Spain to find a direct route to the source of the most valuable spices.

Christopher Columbus came up short when he discovered America. He ended up finding a new route to a new land, one that didn't have the sought-after pepper, nutmeg, mace, cinnamon, or cloves. The Portuguese explorer Vasco de Gama, on the other hand, succeeded beyond all expectations. In fact, not only was he the first to sail around the Cape of Good Hope into the Indian Ocean from Europe, he made it all the way to India and then to Malacca, a major trading port close to the Spice Islands. Da Gama hit the jackpot by finding a marine route to the spices and returning with a spice cargo that more than paid for the whole trip.

The vast wealth flowing into Portugal from its control over the Spice Islands did not go unnoticed or unchallenged by rival European powers. Dutch territorial ambitions in the Indian Ocean were motivated by economic opportunities, particularly the lucrative spice trade with the East Indies. Eventually the Dutch captured all Portuguese possessions in the East Indies and fiercely guarded their monopoly over the production and export of spices for over 150 years until it was broken, thanks to Poivre. After many attempts, he was finally successful in demonstrating that spice plants could be successfully transplanted to other tropical lands, thereby transforming the global market for spices.

Poivre observed the spice trade with interest during his convalescence. He took it upon himself to learn Malay and acquaint himself with Dutch traders who shared their knowledge of their operations with him. Poivre began to think of ways to steal spice plants from the Dutch and transplant them to other tropical colonies under French control, thereby breaking the Dutch monopoly on spices. Cloves, nutmeg, and mace are native to only a few tiny volcanic islands of the Moluccas, now part of Indonesia. The Dutch fiercely guarded these islands, which gave them complete control over the global production of these rare spices.

As a young French invalid recovering from the loss of his arm, Poivre posed no threat to the Dutch traders in Batavia who unwittingly provided him with more useful information than they would otherwise give to a potential adversary. Poivre learned enough from the Dutch to identify weaknesses in their operations and realized that in theory, at least, it was possible to successfully smuggle nutmeg and clove seedlings.

Poivre left Batavia for Pondicherry, then a French colony on the east coast of India, and then to Île de France (Mauritius), en route to France. While on Île de France, he no doubt observed that the climate and conditions would be ideal for transplanting spice plants purloined from the Dutch East Indies.

In 1749 Poivre obtained a royal appointment with the French East India Company and returned to Asia on a number of assignments, always working towards his goal of smuggling clove and nutmeg seedlings from the Spice Islands. He received official backing from the French East India Company for his secret mission and traveled to the Moluccas but was unsuccessful in his efforts to filch clove and nutmeg seedlings from the Dutch. Eventually he was able to obtain nine nutmeg trees through a middleman, which he took back to Mauritius. These plants were too few to reproduce bountifully and were not enough to cultivate into a source of viable spice production.

Poivre was nothing if not determined. In 1755, he made another attempt to penetrate the Dutch-controlled Spice Islands with a new ship but failed to find a sheltered landfall. He also faced an increasingly hostile crew who had not been told of their intended destination and were not amused when he informed them. Poivre headed south and ended up in Portuguese-controlled Timor, only three hundred miles from Australia. There he met a friendly governor who presented him with less than a dozen nutmeg seedlings and a handful of clove seeds as a gift.

Poivre returned to Île de France and planted the nutmeg seedlings in the garden at Pamplemousses. Unfortunately, the plants failed and his relations with local authorities soured. He returned to France in 1756 and on the return voyage, once again, his ship was attacked by the English. He was captured and spent seven months in an Irish jail in Cork before being freed. He used the time to write his memoirs that he published as

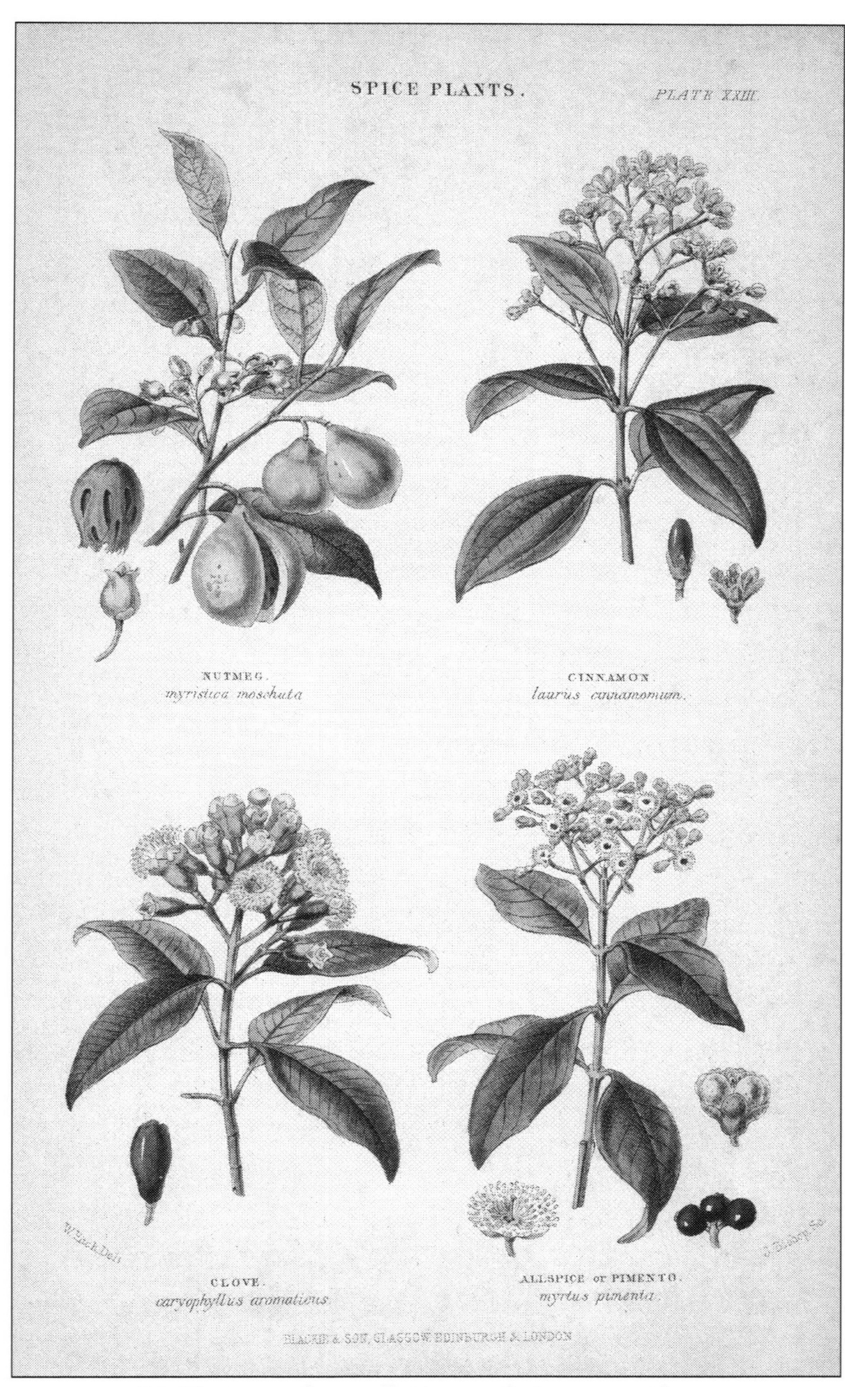

31. Nutmeg, clove, allspice, and cinnamon plants

Voyages d'un philosophe or *Voyages of a Philosopher.* The book describes his travels and observations in Asia. He returned to Lyon, and married Marie-Françoise Robin, with whom he had three daughters.

Meanwhile the French East India Company was dissolved and its assets taken over by the French crown. The French foreign minister, Duc de Praslin, appointed Poivre as Intendant of Île de France and Île Bourbon (La Réunion), and may have been receptive to Poivre's ideas for spice production. The prospect of growing spices on Île de France provided at least a possibility for filling the crown's depleting coffers after having to acquire the assets of the failed French East India Company. Poivre and his family arrived in Île de France (Mauritius) in 1767 to take up his new position.

Poivre was a skilled administrator and helped transform Île de France into a thriving colony. Poivre's contributions as Intendant for Île de France were many. He implemented numerous projects that revitalized Port Louis by clearing the harbor of silt and building water mills, warehouses, and printing works. He established an early hospital in 1770, created the first pharmacy, and in 1772 allowed the first Hindu temple to be built. His most famous legacy, however, was the design and construction of the world-famous botanic garden at Pamplemousses.

The origins of the gardens at Pamplemousses began in 1736 when newly installed Governor Mahé de Labourdonnais bought the property called Mon Plaisir and built a luxurious home there. Labourdonnais created a vegetable garden at Mon Plaisir that was intended to supply not just his household, but provide food for Port Louis and visiting ships as well.

The beginnings of the present botanic garden date from Pierre Poivre's construction of a much larger and more ambitious garden in 1768. In a different location from Labourdonnais' house at Pamplemousses, Poivre built a mansion for himself called "Mon Plaisir" ("My Pleasure" in English) and supervised the design and construction of an elaborate botanic garden. The original Chateau de Mon Plaisir house no longer exists. A duplicate was reconstructed in the middle of the nineteenth century. This building stands today and is used as an office.

He befriended a talented local gardener named Nicolas Céré who shared Poivre's interest in horticulture and together they planted examples of many indigenous species along with many exotic plants and fruits from other lands. Poivre developed a friendship with a clerk named Provost who shared his interest in spices and the two of them planned another clandestine expedition to the Spice Islands to bring back nutmeg and clove seedlings to cultivate in the garden at Pamplemousses.

Poivre supervised the endeavor but did not participate in the voyage himself. Instead, he appointed Provost to lead two French ships first to Manila, and from there to the Spice Islands. The two ships left Mauritius in 1769 for the Spice Islands where they evaded Dutch security and were able to obtain hundreds of nutmeg and clove trees from local islanders. On June 24, 1770, Provost returned to Mauritius and presented Poivre with his cargo of rooted trees along with hundreds of seeds and nuts. Cloak-and-dagger espionage has a long and colorful history in Mauritius.

The smuggled trees were planted at Montplaisir and the first trees bore fruit in 1775. Cloves were later planted at other tropical French colonies including Réunion, Seychelles, French Guiana, and Martinique. After 1818, cloves from Mauritius were transplanted to Madagascar and Zanzibar, now the most important source of cloves in the world. Poivre had successfully broken the Dutch monopoly on cloves and nutmeg, forever changing the spice trade and influencing the balance of European power in Asia.

Poivre's legacy stretches all the way to the United States. He returned to France and died in 1786. His wealthy widow married again to Pierre Samuel du Pont Nemours, an advisor to the King of France. Not a good job to have during the French Revolution. They emigrated in 1799 to the United States upon the encouragement of Thomas Jefferson, whom they had befriended while he was the U.S. Minister to France. Pont de Nemours' son founded E.I. du Pont de Nemours and Company, one of the largest industrial empires in the United States.

In Mauritius, Poivre's most accessible and enduring legacy is Pamplemousses, even though his name or story isn't obvious to the casual SSR Garden visitor, except on the avenue named after him and a bust of himself on a stone pedestal. The visual beauty of the garden is certainly a sight to behold especially if you are interested in plants. For me, the smells of the garden were more of a surprise. There are many aromatic plants throughout the garden including ylang ylang (pronounced as ee-lang, ee-lang). This tropical flower is grown both on Mauritius and La Réunion for its aromatic oils, which are used to make perfume.

In addition to a huge variety of flora, the garden also has native fauna on display including the Chauve-Souris, a flying fox (*Pteropus niger*). I saw dozens of giant bats flying among the fruit trees. These large bats seemed like loud dogs with wings, something I didn't want to examine closely.

The SSR garden has great national significance for Mauritius and many important foreign visitors have made the trek to Pamplemousses. The garden is filled with historic monuments and has one section filled with a row of trees dedicated to special state occasions. Each tree was planted by a visiting foreign dignitary and is marked with a commemorative plaque. The names on the plaques represented an interesting mix of royalty and world political leaders. I noticed plaques with the names of Indira Gandhi, Nelson Mandela, HRH Princess

Margaret, HRH Princess Anne, and the inimitable Robert Mugabe of Zimbabwe.

It has taken me several visits to appreciate the full scope of local history that the SSR garden represents in a single place in Mauritius. The large formal garden from the eighteenth century is a surprising legacy of French power and influence in this part of the world and the significance and story of the spice trade is an important if forgotten piece of world history. The garden itself displays the wide range of native plants found in the Mascarenes in addition to introduced varieties that can grow here. The more I saw and learned in the garden at Pamplemousses, the more I wanted to see the other major island of Mauritius.

CHAPTER 9

RODRIGUES ISLAND

DURING SEVERAL VISITS TO MAURITIUS, I heard quite a bit about Rodrigues Island. It was time to check it out for myself. It's the only populated region of Mauritius not connected to the main island and comes up in conversation frequently. I never heard much other than "To see all of Mauritius, you really must go to Rodrigues Island." Getting to Rodrigues takes some planning, however.

Transportation options to Rodrigues are very limited with only a few daily flights from Mauritius on a 44-passenger prop plane. The remote island lies out in the middle of the Indian Ocean, 372 miles (600 km) to the northeast of Mauritius. The nearest landfall to the east of Rodrigues lies thousands of miles away in

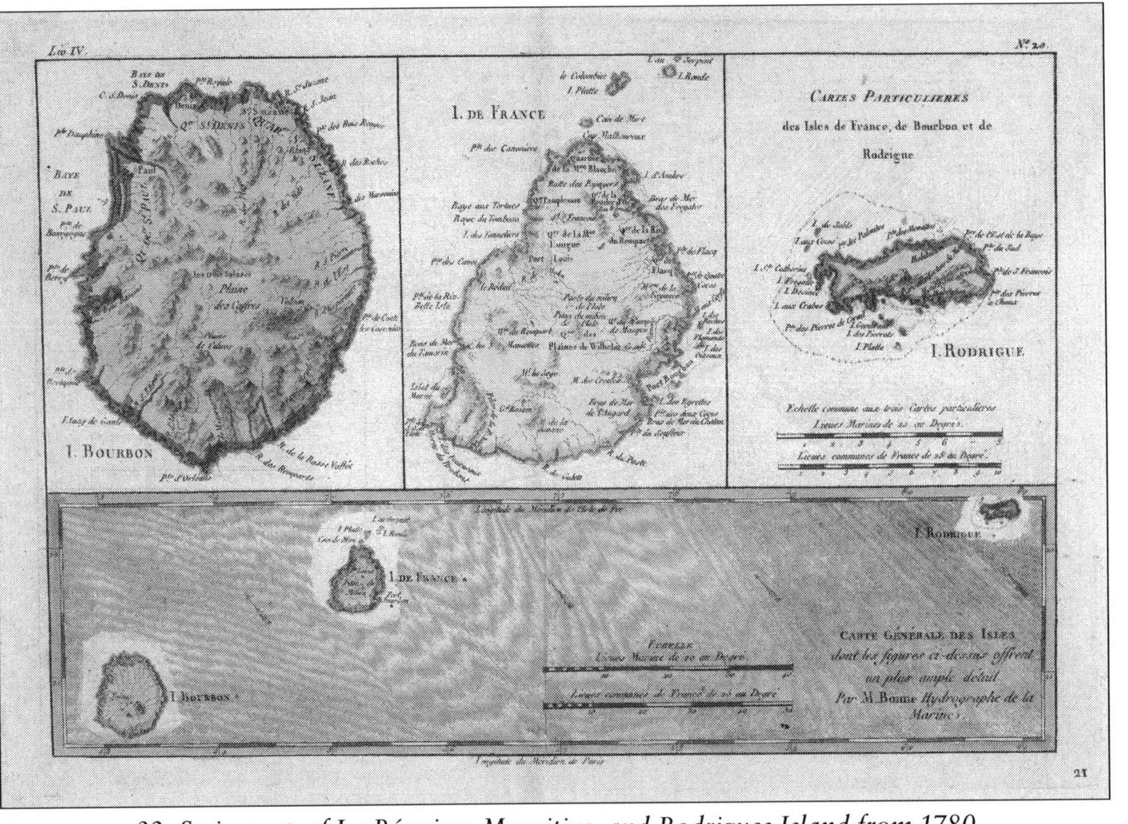

32. *Swiss map of La Réunion, Mauritius, and Rodrigues Island from 1780*

Western Australia. The flight takes about an hour and a half, although my flight returned to Mauritius just 30 minutes after takeoff to pick up a mechanic and some parts to take to Rodrigues. This delayed my arrival and set my nerves on edge. Apparently this sort of delay is very common and I discovered that Rodrigues runs on its own unhurried island time.

When I finally arrived on the tiny island that is only eight km wide and 18 km long (roughly five miles wide and 11 miles long), my hosts were unconcerned about my delay. As we were heading back to their home, they gave me an impromptu sightseeing tour and asked if I wanted to go dancing. I laughed and said I wasn't quite ready to switch gears that quickly. They told me that on Sunday afternoons either you drink or go dancing for entertainment on Rodrigues Island. And sure enough, as we passed Les Cocotiers Night Club, people were streaming toward the entrance.

Rodrigues has only about 37,000 residents, the vast majority of whom are African Creoles. It is a stark contrast to the large mélange mix of races and religions on Mauritius. On Rodrigues, most of the population is Roman Catholic and the church calendar plays a big role in the island society. Economically, Rodrigues is the poorest part of Mauritius, dependent mostly on local fishing and agriculture.

Tourism is still in its infancy. Formal tourist attractions per se are almost nonexistent and the main reason

33. Saint Gabriel Church, Rodrigues Island

to visit Rodrigues Island is to enjoy its natural scenic splendor. There is only one gas station on the island and a handful of small stores. The most impressive man-made structure I visited was the St. Gabriel Church (Eglise St. Gabriel). This very large and open-air church was built by hand of volcanic rocks, a painstaking effort by the many deeply religious Creole parishioners living on the island. It is a beautiful and unusual church, complete with birds nesting in the rafters.

Accommodations consist of a few small hotels. Most visitors stay in small guest houses *(chambres d'hôtes)* that in my opinion are the best way to experience Rodrigues. People on Rodrigues are very warm and welcoming, and staying in a guest house is a wonderful opportunity to learn about the island and taste some delicious food in the process.

Rodrigues is unlike any place I have seen. The beauty is fragile and ethereal. The white sand beaches are stunning and empty. The enormous coral reefs surrounding the island give the waters a distinct aquamarine hue. While the perimeter of the island gives the impression of a dry, often rocky island, the hilly interior is topped with green growth and patches of tropical forest. Despite the verdant interior, fresh water is in short supply on the island and strict water conservation measures are always in effect.

The remote island has been described as "the best kept secret in the Indian Ocean." I think this is meant to describe its beautiful beaches and amazing lagoon that are perfect for diving and snorkeling. I don't think anyone is trying to keep it a secret however. The tourist infrastructure, such as it is, is very small and the government is trying to develop tourism very carefully so as not to overwhelm the small island's fragile ecosystem. Rodrigues is not set up for luxury vacations.

Being so isolated, the Rodrigues Islanders are a resourceful lot. They have to be! Port Mathurin, the largest town and de facto capital of the island, has a population of 4,000 and feels like a small village. Other communities on the island are tiny, connected by a public bus service that runs on paved but often very bumpy roads. The bulk of the food is produced locally. Very few if any pesticides are used anywhere making nearly all of the produce and livestock raised on the island naturally organic. Local seafood is also pristine because it comes from an extremely remote location in the Indian Ocean far from sources of industrial pollution.

Octopus is plentiful in Rodrigues and historically has been an important food source and appears in many local dishes. I was asked enthusiastically by one resident, "Did you know that you can use coat hangers to dry an octopus in the sun?" No, I didn't know that. Why I've been using them just to hang up my clothes all these years is obviously a mystery. After this erudite observation, while driving around the island I did notice lots of octopuses drying in the sun and many were strung up on coat hangers.

There is one abattoir (slaughterhouse) on the island that is used for cattle, goats, and pigs. Tethered cattle are left to graze by the roadside and it seems like "free range" goats and chickens wander everywhere on the island. There are no predators to threaten them. The only native mammal left on the island is the extremely endangered Rodrigues fruit bat *(Pteropus rodricensis)*. Deforestation destroyed the bats' natural habitat plus periodic cyclones put additional pressure on the bats. Captive breeding and other conservation programs have helped to restore their numbers in recent years. I saw a number of these bats flying around during my short

34. Tethered cattle are a common sight on Rodrigues

stay on the island, so while threatened they do seem to be alive and well.

All three of the Mascarene Islands suffered from the arrival of people. With its small land area and lack of high mountains, Rodrigues was easy for humans to overwhelm. Before the first Europeans arrived 400 years ago, it was a lush green island throughout. Forests of native palms and other trees covered the entire area with plenty of water supplied by numerous small rivers. Over-hunting for tortoises, logging to clear land for farms and fuel, fires, and introduced animals turned much of the island into a parched patch exposed to the intense tropical sun. Destruction of native habitat for many endemic species, especially birds and lizards, hastened their extinction on such a small island.

Not all is lost. There are several conservation projects on the island attempting to protect and restore the natural wildlife that has survived. The most illuminating place I visited on Rodrigues was the François Leguat Giant Tortoise and Cave Reserve. This 47,000 acre preserve offers visitors a chance to examine local geological history in the unique limestone caves found on the island, see over a thousand giant Aldabra tortoises (*Dipsolchelys elphantina*) wandering about, and view the successful planting of more than 100,000 indigenous trees in an effort to recreate the natural environment of the island before man.

These introduced tortoises substitute for the extinct Rodrigues Giant Tortoise that once numbered in the hundreds of thousands. The Rodrigues Giant Tortoise *(Cylindraspis vosmaeri)*, a giraffe-necked species of giant tortoise native to Rodrigues Island, was hunted to extinction by 1802. Crews of sailing ships harvested these tortoises by the thousands to supply fresh food and act as living ballast.

The small François Leguat Museum on the reserve has fascinating exhibits describing the history and the flora and fauna of the island. I found the small museum memorable because it introduced me to two famous former inhabitants of Rodrigues Island that exemplify the intertwined fate of man and nature. At this museum I was introduced for the first time to the extinct Rodrigues Solitaire bird and the only Western chronicler of this anti-social animal—the late explorer and adventurer extraordinaire François Leguat de la Fougère.

The Rodrigues Solitaire *(Pezophaps solitaria)* was the "dodo" of Rodrigues Island, a large flightless bird that evolved from the same Nicobar pigeon ancestor of the Mauritius dodo. It adapted perfectly to its environment on Rodrigues where it had no predators, lost its ability to fly, grew to a height of about one meter, and weighed nearly fifty lbs. Called a Solitaire (or solitary) because it was often found alone, this bird also became extinct for the same reasons as the Mauritius dodo. It was hunted by man for food, and became prey for introduced species like rats, cats, dogs, and pigs. François Leguat first noted the Solitaire when he was stranded on the island for two years beginning in 1691. Sightings of these birds became more infrequent over the years and they are believed to have become extinct around 1750.

Just as the Mauritius dodo has become an iconic symbol of Mauritius, the Solitaire has become an icon for Rodrigues Island where its image appears everywhere. There is a large statue of a Solitaire on a pedestal in Port Mathurin. The bird figures prominently in local advertising, postcards, and in all manner of gift shop souvenirs. Only one realistic picture of a Solitaire by an eyewitness exists, and it was drawn by François Leguat. His drawing depicts a bird that is more elegant in appearance than the Mauritius dodo with a long, slender neck and smaller beak. Unlike the dodo, the Solitaire did not travel well and never made it to Europe. It was not immortalized in great works of fiction. And much like Rodrigues Island itself, it remains unknown to the larger world.

As I explored more of Rodrigues Island, I discovered how much of its early history was interwoven with the fate of Leguat. Who was this Leguat? Leguat was not the first European to visit Rodrigues Island but he was the first to live there and document his observations. As with the case of Pierre Poivre, I discovered that Leguat's life was filled with amazing adventures and accomplishments.

Rodrigues Island had been visited by early Arab mariners prior to the arrival of Europeans. It was discovered in 1528 by Diego Rodriguez, a Portuguese mariner who named the island after himself. Some historians maintain that it was discovered in 1507 by another

1822.—Le Solitaire, from Leguat.

35. Drawing of Rodrigues Solitaire by François Leguat

Portuguese explorer, Diego Fernandez Pereira. But the fact is no one lived on the island for close to 200 years after its discovery. The Dutch largely ignored Rodrigues and it was a group of French settlers led by Leguat who colonized the island in 1691. For Leguat, Rodrigues was just one segment of his seven-year saga of harrowing hardship and survival that he turned into a best-selling book when he finally returned home to Europe.

Leguat was a French Huguenot farmer from the town of Bresse in France who acted as the leader of a group of eight men that landed on Rodrigues in 1691. Huguenots were French Protestants, driven into exile or massacred in 1685 by King Louis IV, who revoked the Edict of Nantes making it a crime to be a Protestant in France. Leguat escaped from France and settled in the Netherlands in 1689 where he and some friends learned of a plan to establish a colony of French Protestant refugees on a Dutch-controlled island in the Indian Ocean. The French Marquis Henri du Quesnes organized the project under the authority of the Dutch East India Company and the Dutch government. Leguat and his companions were attracted to the idea of establishing a utopian Huguenot settlement on a tropical island.

Their plan was to establish a colony on Île de la Réunion, then known as Île Bourbon, which they thought had been abandoned by the French at that time. Leguat and a group of followers signed on to the expedition that left from Amsterdam on the ship Hirondelle on July 10, 1690 ostensibly headed for Île Bourbon. However, before the Hirondelle left Amsterdam, the Marquis learned that Île Bourbon had in fact been re-annexed to the French East India Company, which would prevent

36. The prison on Rodrigues Island is decorated with a Solitaire

Huguenots from settling there, but chose not to tell Leguat and his men.

The captain of the Hirondelle was instructed not to deliver them to Île Bourbon but to Rodrigues Island instead. Leguat and his men were left on Rodrigues and told that it was just as good as Île Bourbon. They quickly realized they had been marooned and it took them two years before they were able to escape to Mauritius by building and sailing on a raft. When Leguat later learned of the deception, he accused the Hirondelle captain of "the basest treachery." I'm sure that he would have used stronger language in today's vernacular.

Fortunately, Rodrigues was a lush island with plenty of food so the castaways were able to survive. During this time Leguat documented his observations of the abundant flora and fauna, much of which later became extinct. Leguat's extensive written descriptions and hand drawings of Rodrigues are today an invaluable scientific record of the species that once lived on the island.

After two years, Leguat and his companions gave up hope for the arrival of any ships bringing them more Huguenot settlers from France. The sense of abandonment (and according to other observers, a lack of female companionship) persuaded Leguat and his companions to find a way to escape from the island. With great resourcefulness, they built a raft and made a harrowing voyage across hundreds of miles of unprotected ocean to Mauritius.

Unfortunately, instead of finding freedom on Mauritius, they were imprisoned by the Dutch governor there as enemy aliens. After three years of incarceration, they were sent to the Dutch colony of Batavia (modern Jakarta, Indonesia) where they remained prisoners until a Dutch Council established their innocence.

In March 1698, Leguat and two surviving members of the original Huguenot expedition to Rodrigues returned to the Netherlands. He later moved to England and wrote a book about his travels, *A New Voyage to the East Indies,* which was published in 1708 and became a best seller. His drawing of the Solitaire was included in this

37. One of many beautiful beaches on Rodrigues Island

book. For many years, Leguat's published observations about the native flora and fauna were ridiculed by some scientists and even dismissed as fiction by others.

Since then, scientific analysis of bird bones and other material discovered in caves on Rodrigues Island confirmed the accuracy of his findings, including his description of the Rodrigues Solitaire. The illustration he made of the Solitaire is the only surviving image of this bird made by an eyewitness. The book remains an important work for the extensive and accurate descriptions of now long-extinct wildlife that Leguat encountered during his stay on Rodrigues. No wonder the nature preserve and museum is named after him.

After learning about Leguat's long and adventurous life, I am in awe not just of his scientific accomplishments but am also impressed by his amazing physical constitution. He was over fifty at the time he left on his voyage to the Mascarenes. After surviving two years on Rodrigues Island, then enduring years of imprisonment, he returned to Europe at age sixty, sailing across the world under the difficult conditions of sea travel at the time. He then wrote a best-selling book, and is believed to have died at the age of ninety-six in 1735. Reaching this age is a major achievement today but was downright amazing during the eighteenth century when life expectancy was only about half of what it is today.

Rodrigues Island is very much off the beaten path and is well worth the trip. It's one of the most unspoiled places that I've seen. Compared to the U.S., it's refreshingly uncommercial. In fact there aren't many stores at all, only a handful of small shops. Rodrigues is not economically rich but has a deep sense of community and identity. The airport is modern but very small and can handle only prop planes. The runway seems a tad short for the forty-four-passenger Air Mauritius plane that normally lands there.

For many people I encountered, I was the first American they had ever met. I discovered one of the world's slowest Internet connections when trying to use my laptop on the public library in Port Mathurin. However, plans are in the works to upgrade connections to the outside world. The airport will be expanded to accommodate larger aircraft and a fiber optic cable to

Mauritius should provide high speed Internet access to the island in a few years.

Let's hope the island is not spoiled further. The environment now, while beautiful, is in sharp contrast to the lush forested gem that once existed. The verdant paradise that was home to the Rodrigues Solitaire is gone, thanks to humans. The Rodrigues Solitaire became extinct probably only sixty years after Leguat and his companions arrived on the island in 1691. The abundant streams and forests have vanished, though there are efforts underway to restore some of the ecological health back to the island.

Rodrigues has taken much longer to develop than the rest of Mauritius, and during this time much has been learned in terms of environmental conservation techniques and the importance of ecotourism as a sustainable economic policy. Fortunately, tourism development in Rodrigues is proceeding cautiously, so as to protect the island's threatened endemic species and fragile ecosystem.

The quick disposal of other species by man is disconcerting. The Mauritius dodo was probably extinct by 1690, less than one hundred years after the Dutch first arrived on Mauritius in 1598. The sequence of events leading to the dodo's extinction, the Solitaire's demise, the tortoise's extermination, and the skink's disappearance from the Mascarenes has been tragically duplicated in other parts of the world.

It is a sad tale to keep repeating.

CHAPTER 10

EXTINCTION

Even if I didn't know that Mauritius was home of the dodo, there was no escaping the strange and comical bird's image once I arrived. There were dodos everywhere! Dodos in shops, dodos in restaurants, dodos featured in museums and other places that tourists are likely to visit. The bird may be extinct in real life but it is thriving in Mauritius advertising.

The dodo is emblazoned on coffee mugs and beach towels, and all manner of tourist souvenir gifts, and you can even bring it home with you. With all the promotion, it's easy for foreign visitors to think that the dodo is the national symbol of Mauritius.

38. *A dodo identified as "Didus Ineptus,"* 1806

But I wondered how the Mauritians perceive their national icon. In conversation, I would often ask residents how they felt about the dodo. Most people laughed and seemed to regard the dodo as a long-dead local bird that is now a marketing gimmick. More thoughtful answers pointed to the negative connotations of the dodo for Mauritius. One gentleman told me, "When we think of the dodo, we remember it with sadness because it is gone forever."

Dead as a Dodo.

There is a reason that phrase comes up when something is beyond hope, gone forever, and unequivocally dead. Strange that an odd bird, which only coexisted with man for less than a hundred years, virtually lives on centuries later. But then again, we need the dodo to explain how our world is changing.

Today species are dying out at a faster rate than any other time save for the dinosaurs' demise 65 million years ago. The dodo was one of our initial alarm bells. After the complete disappearance of the dodo around 1690, the dots were connected between the activities of man and a species extinction. The iconic dodo represents the recognition that the world's natural resources are not limitless and that human intervention with nature can have tragic, irreversible consequences.

Why does the dodo still haunt us, I wondered, when on Mauritius alone there were dozen of birds species lost. Yet, those ducks, parrots, and owls are forgotten. Maybe

39. "Dodo Presenting a Thimble" from Alice's Adventures in Wonderland, *1865.*

what troubles us is the fact the trusting dodo actually went up to the human visitors, who then slaughtered the naive bird. Yet this bird was hardy enough to make the six-month journey to Europe. This proves that the species could have survived if just showed some care. Unlike many extinct species, dodos were sturdy creatures and I can't help thinking that if only a few early settlers took it upon themselves to protect just a small breeding population of these birds, if only for curious collectors in Europe, the dodo would still be with us today.

Maybe the dodo stands out because it was so bizarre looking. Some of the birds that made it to Europe were captured in paintings and taxidermy. This art catapulted the dodo into the collective consciousness. It was the remains of a dodo in a museum that inspired Lewis Carroll to include the dodo in his classic novel *Alice's Adventures in Wonderland*. Thanks to that popular tale, the dodo remains preserved in our imagination today.

And you can't help but remember the eccentric bird if you are on Mauritius. If the dodo is considered to be Mauritius' national symbol, it is not exactly an upbeat one. Many Mauritians would prefer a more vibrant and still existing bird for a national symbol, namely the Mauritian Kestrel, the Pink Pigeon, or the Echo Parakeet. The kestrel is the symbol used on Air Mauritius airliners, and individual airplanes have names that include Pink Pigeon and Parakeet. I can certainly understand why no one wants to use a flightless, extinct dodo as a mascot for the national airline. It doesn't exactly inspire confidence.

The dodo and other Mauritius birds are the products of a rare evolutionary path. By virtue of chance and their remote location, the Mascarenes were left undisturbed by humans until the sixteenth century. For millions of years Mauritius was a veritable paradise for the dodo where it followed a unique, unobstructed evolutionary path. Over time the dodo evolved from a flying pigeon into a bizarre bird with a huge body, tiny flightless wings, and a strange bald face with a giant beak. Despite this strange appearance, the dodo was perfectly adapted for life on Mauritius. It could store fat in its large body to survive during lean times. Its bulbous beak was perfect for cracking seeds and nuts. And with no predators, the dodo lost the need for flight.

These evolutionary adaptations presented no problems as long as Mauritius remained free from predators. Once humans arrived in the sixteenth century, the dodo faced a continuing series of lethal threats that sent it spiraling towards extinction. Destruction of natural habitat, introduction of predators, and hunting put not only the dodo but many islands' species on a one-way path to oblivion.

Beginning with the Portuguese in 1507, European visitors to Mauritius hunted the dodo and other native animals for food. With no fear of humans, the Dodos were easy prey and were slaughtered by the thousands.

These first visitors to Mauritius were hungry sailors who had been at sea for months on minimal rations and found the large goose-size birds easy to catch. Other endemic species had no fear of man making them similarly easy to capture, even those that could fly. Records from a 1607 Dutch voyage to Mauritius describe sailors capturing turtledoves by hand.

The introduction of non-native animals proved more lethal to the dodo and other endemic species than did hunting by humans. The Portuguese did not settle on Mauritius but they used the island as a resting place and a food source. They introduced domestic animals hoping that they would multiply in the wild and provide meat for future visits. In addition to goats, chickens, and pigs, the Portuguese probably introduced rats by accident.

Rats and pigs multiplied quickly in great numbers and were especially deadly to dodos. The dodo laid a single egg in an open ground nest that was an easy target as were defenseless dodo chicks. These predators quickly overwhelmed the ability of the dodo to reproduce fast enough to replace their dwindling numbers. The arrival of the Dutch in 1598 began a period of frequent contact that fostered over-hunting, increased habitat destruction, and deforestation, and sparked an explosion in the number of introduced animals.

Among the introduced predators that decimated the dodo were monkeys that could easily grab eggs and defenseless chicks. No one knows who introduced monkeys to Mauritius, a species known as crab-eating macaques, but they reproduced quickly and still live on the island today.

I've seen these monkeys up close at Ganga Talao, the Hindu temple complex also known as Grand Bassin in the south of Mauritius. The monkeys live in the adjacent woods and have learned that visiting crowds to Ganga Talao are a sure bet for a free meal. The macaques are very small monkeys that cling together in groups. They are tame enough to come up to people and grab a piece of food or candy, then scurry off to eat it in safety. The monkeys are cute but they are still wild animals with very sharp teeth.

In 1662, after the Dutch abandoned their first attempt at permanent settlement in 1658 and before they tried again in 1664, a Dutch sailor named Volquard Iversen was shipwrecked on Mauritius. He and his fellow survivors searched for food and found dodos on a small islet off the coast, but not on mainland Mauritius. The dodos probably survived here because the sea protected the islet from pigs, rats, and monkeys that had overrun the mainland by that time. Iversen and his shipmates captured some of the birds and cooked them over a fire as food. After they treated themselves to this dodo barbeque, a passing ship rescued them just five days later. Iversen's was the last known eyewitness account of a living dodo.

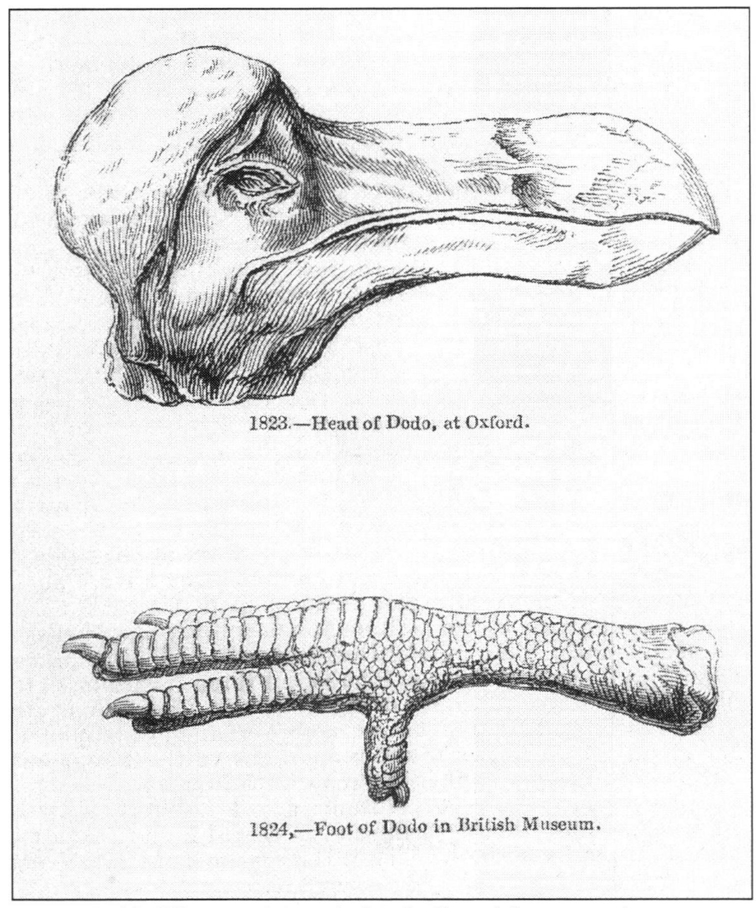

40. The surviving dodo skull and foot remains

One of the last known dodos in Europe was a taxidermy specimen that belonged to a wealthy English collector named John Tradescant who died in 1638. Tradescant was a wealthy collector and naturalist who had a collection of curiosities that he displayed to the public for a fee at his house in Lambeth. His collection formed the basis of the Ashomolean Museum in Oxford and later became part of the Oxford University Museum of Natural History. This particular dodo was most likely one of the few live birds shipped to Europe from Mauritius in the early seventeenth century and preserved through taxidermy after its death.

By 1755, this overly ripe dodo specimen was probably more than a century old and decomposing. Legend has it that orders were given to burn it as trash, yet some enterprising individual thought to remove the dodo's head and foot, saving them from destruction. A good tale but dubious. This story of saving the carcass from a fire is disputed and it may be that the head and foot were the only parts of the body not to decompose.

It was here in the Oxford natural history museum that Charles Lutwidge Dodgson a.k.a. Lewis Carroll encountered the dodo. Dodgson was a mathematics lecturer at Christ Church College in Oxford who befriended Alice Liddell and her sisters Edith and Lorina, the young children of the Dean of Christ Church. He took Alice and her sisters on visits to the museum where the dodo was one of Dodgson's favorite exhibits. A painting of the dodo by Jan Savery hangs in the museum and is thought by some to be the original inspiration for the dodo character in *Alice's Adventures in Wonderland,* published in 1865.

41. "Mouse Telling a Story" from Alice's Adventures in Wonderland, 1865

This book developed from the stories that Dodgson created to entertain Alice and her sisters based on places, people, and things they saw on their outings together. The dodo character in the book is thought to be a self-deprecating caricature of Dodgson himself. He had a stammer and often inadvertently introduced himself as "Do-do-Dodgson."

It is interesting to trace the etymology of the word dodo since I believe part of the bird's allure is the catchy name. There are literally dozens of different names that have been used to describe the dodo since its discovery on Mauritius. However, none of them are very flattering, not even the scientific ones. The dodo was so odd

42. *Dodo, cassowary, ostrich, and penguins, 1790*

looking that the Portuguese, the first Europeans to reach Mauritius, described it as "Doudo, " which in sixteenth century Portuguese is "crazy."

The first Dutch explorers to see the bird called it "walghvogel" which means disgusting bird or loathsome bird because it tasted bad. That didn't stop the Dutch from hunting and eating these birds in large numbers. The Dutch word "dodoor" refers to sluggard, which might have been used to describe these fat, slow-moving, and flightless birds. Another theory suggests that the term dodo originated from the Dutch word "dodaars" which in English means knot arse and refers to the knot of feathers on the rear end of the bird. Later Dutch expeditions called the birds by different names including "doederssen," "dronte," and "dronten."

The first naturalist to give a scientific description to the Mauritius Dodo was Carolus Clusius of France who in 1605 published observations of the remains of a specimen. Clusius gave the dodo a Latin name *Gallus gallinaceus pergrinus* which roughly translated means foreign cock of the chicken family.

It wasn't until 1628 that the term "dodo" first appears in the English language to refer to this curious flightless bird of Mauritius. The poor bird didn't fare much better on the science side. *Raphus cucullatus* is the scientific name originally given to the dodo by the Swedish Botanist Carolus Linnaeus in 1758. The approximate translation for this name is "cuckoo-like bird with a fat rump." Later in 1766, Linnaeus changed this name to *Didus ineptus*, which means clumsy dodo. The *Didus ineptus* name later fell out of favor and the scientific name reverted to *Raphus cucullatus*, the term still used today.

By the early nineteenth century, the dodo had become completely forgotten by most Mauritians and some asserted that it was a mythical bird that never existed. Scientific interest in England was rekindled with the publication of several works about the dodo. A curator named John Theodore Reinhardt in the early 1840s examined the head of a dodo that had been in the collection of the Royal Museum in Copenhagen for nearly 200 years. Reinhardt was the first to suggest that the dodo was in fact a very large, flightless pigeon.

In 1848 Hugh Strickland and A. G. Melville published a book called *The Dodo and its Kindred* that, among other things, asserted that the dodo had in fact been a real bird and that it was the first definitive example of a species extinction caused by man. Referring to the Mauritius dodo and the Rodrigues Solitaire, the authors state that these birds "furnish the first clearly attested instances of the extinction of organic species through human agency." They also concurred with the hypothesis that the dodo was a member of the pigeon family.

This was a period of renewed study in the natural world, a movement towards observational science that culminated in Charles Darwin's publication of his book *The Origin of Species* in 1859. As a young man, Darwin visited Mauritius in April 1836 while on the final leg of his five-year global voyage. Darwin made no mention of the dodo in his descriptions of his visit to Mauritius.

Darwin may have ignored the dodo, but the recognition that a species disappeared due solely to the intervention of man was a monumental event in scientific thought and in public awareness. For the first time in history, humans became aware that the earth's resources were finite. This in turn led to a concept of preservation of species, a fundamental goal for modern environmental conservation.

The same year *Alice's Adventures in Wonderland* was published, another major milestone in the study of the dodo occurred. In 1865 George Clark, a British school teacher living in Mauritius, was alerted by a railroad engineer to the discovery of a cache of dodo bones in a swampy area called Mare aux Songes, near the site of today's international airport. Today the excavation site, marked by an inauspicious statue of a dodo behind a chain link fence, lies close to the airport's runways.

Clark had an abiding interest in natural history and had searched unsuccessfully for years to find dodo bones or some other physical evidence on Mauritius. He went to the railroad construction site where the bones were found and began digging. Clark excavated hundreds of dodo bones, many of which he sent to the British Museum and other major museums around the world for study.

A number of dodo skeletons were reassembled from Clark's diggings and other excavations. Additional discoveries on Mauritius over the years have produced many more dodo bones, providing physical evidence establishing the fact that the dodo was in fact a very real bird. The fossil record remains incomplete however, as no dodo eggs or embryos have yet been found.

When it comes to scientific knowledge, I find it both surprising and puzzling that more is known about dinosaurs than the dodo. The Ashmolean Museum once again plays an important role in preserving the heritage of the dodo. The fact that two key pieces (that same salvaged head and foot!) of a dodo's body with soft tissue survived all this time in the museum allowed scientists to recently conduct DNA analysis. The analysis confirmed the dodo was related to the pigeon.

Despite its sketchy fossil record and short appearance during recorded human history, the dodo remains one of the most famous vanishing acts. The bird itself has become synonymous with extinction. Though many people can't find Mauritius on a map, nearly everyone knows the expression "Dead as a Dodo" means permanently and forever dead.

While the dodo remains irrevocably lost, the island that gave birth to this oddity endures. And there are efforts being made to preserve the remaining unique species. One of the most amazing things about Mauritius is that the site of history's most famous extinction now promotes some of the most successful conservation programs in the world.

CHAPTER 11

SURVIVAL

LONG AFTER THE DODO'S EXTINCTION, the causes behind its demise continued unabated for hundreds of years. Each successive wave of European administration of the island brought its own set of environmental problems, compounding the destruction of species that thrived before human activity and predators.

Deforestation began under Dutch occupation but was focused mostly on harvesting the rare endemic ebony trees for piano keys. It was under French administration beginning in 1721 that dramatic deforestation occurred as sugarcane production increased. Much of the island's lowland forest was cleared to plant sugarcane. The sugar was then processed by steam power mills fueled

43. Engraving of a dodo from 1790, long after extinction

by locally cut firewood. The logging further reduced forests, increasing pressure on wildlife by shrinking their habitat even more.

Surviving birds and other animals retreated to the more remote areas of the island where their natural environs remained intact. Although many of these areas in remote canyons and mountainous areas were effectively inaccessible to humans, they were no defense against nimble and agile introduced predators especially rats, monkeys, and the mongoose.

During the twentieth century, an invisible but deadly threat to native wildlife appeared in the form of pesticides including DDT, which was widely used across Mauritius from 1948 to 1973. DDT is especially hazardous to birds because it enters the food chain and makes it difficult for eggs to hatch.

All of these factors combined with a rapidly growing human population put enormous pressure on surviving endemic species. Faced with such daunting prospects for extinction, I find it amazing that so many native species were able to hang on at all.

Fortunately, beginning in the 1970s, the picture began to brighten as conservation organizations began implementing pioneering programs to halt species decline and restore native habitats. At that time, the biodiversity of Mauritius was the third most threatened in the world after the Galapagos and Hawaii, with a number of endemic species on the very brink of extinction. Thanks to the efforts of the Mauritius Wildlife Foundation (MWF) and other groups, Mauritius has saved many endangered species and has become a leader in wildlife recovery programs. These include both restorations of native habitats and species management programs.

The Mauritian Kestrel is perhaps the most dramatic example of rescuing a species from extinction. This peregrine-like bird of prey was reduced to a dangerously tiny population in the wild, thought to be only four individual birds by 1971. Beginning in 1973, the Peregrine Fund was instrumental in developing a program for the kestrel that utilized captive breeding and wild management of these birds. The remarkable success of this approach dramatically increased the number of Mauritian Kestrels in the wild which today is close to a thousand birds.

Similar intensive management techniques have been successfully applied to programs to save other threatened endemic birds in Mauritius including the Echo Parakeet, the Pink Pigeon, and the Mauritius Fody. The goal of these efforts is to restore endangered species while simultaneously restoring threatened ecosystems and native habitats. Mauritius and the other Mascerene islands have a high percentage of endemic plants and many of these are declining or are seriously threatened. Many of these plants have been successfully propagated in nurseries then reintroduced to their native habitats.

44. View of Mahébourg Bay from Île aux Aigrettes

One of the leaders in habitat restoration, The Mauritius Wildlife Appeal Fund (MWAF) was formed in 1984 to raise funds and coordinate conservation work being conducted by different international conservation organizations. In 1986, the MWAF undertook its first project, the long-term lease and management of Île aux Aigrettes.

This is a low coral island off the south coast of Mauritius near Mahebourg. The goal of the MWAF was to restore the island to its natural state before the arrival of man. It began by removing invasive species and replacing missing flora and fauna. A secondary goal was to establish a site to propagate and grow endangered endemic plants. In 1994, the MWAF became the Mauritian Wildlife Foundation (MWF) and has since established many other projects.

Île aux Aigrettes has many advantages as a location for a nature reserve, where plants are relatively safe and there are no browsing mammals like rats. The island has the largest remnant of the original Mauritian dry coastal ebony forest, and also has many native plants. The MWF operates a number of conservation programs on Mauritius, but Île aux Aigrettes is the only site open to the public, with guided tours available only by advance reservations.

For visitors interested in the natural history of Mauritius, a guided tour of Île aux Aigrettes is not to be missed. It's probably the single best way to see what the wildlife of Mauritius looked like before the arrival of humans at the end of the sixteenth century. A visit also provides a great opportunity to observe endemic species in their natural habitat, and to learn about the MWF and its successful conservation programs.

The Mauritius Pink Pigeon is one of the rarest birds in the world and Île aux Aigrettes is probably the only place to easily see one in the wild. During my tour of the island, I was able to get close enough to a Pink Pigeon to see that it is indeed very pink in color, as if it were a white pigeon dipped in a vat of pink dye. Giant tortoises imported from Aldabra Island have been introduced to

Île aux Aigrettes to fill the ecological niche vacated by the extinction of the endemic Mauritius giant tortoise. The tortoises have been effective in distributing seeds as they roam about the island. This way, the Aldabra tortoises reseed and propagate native forests in the same way the extinct Mauritian tortoises did before humans arrived on Mauritius.

The museum on Île aux Aigrettes is small but was especially interesting both for its fossil samples and life-size reproductions of some original species that became extinct with the dodo. I noticed a cross section of an ebony tree log that was about one foot in diameter, about six inches thick, and had a solid dark ebony inner circle about four inches diameter. It looked like a giant pencil with the ebony in the same place as a pencil lead.

Without telling me what it was, a museum guide asked me how old I thought this piece of wood was and I guessed about 150 years. I was shocked to learn it was more than 500 years old when it was cut down. The senseless destruction of a rare forest seems even more pointless when you realize how slowly these trees grow.

It was during the tour at Île aux Aigrettes that I was introduced to some of the other endemic species that became extinct along with the dodo. The museum had life-size reproductions of these on display including a Giant Skink and a huge parrot. Known as the Broad-billed Parrot or Large Raven Parrot, this looked like a very large macaw with a giant beak and has the scientific name of *Lophopsittacus mauritianus*. Little is known about these birds except from fossils and accounts from early European visitors who sometimes called them Indian Ravens.

They are sometimes described as flightless but the giant parrot could fly, just not very well. This made them vulnerable to hunting and to introduced predators like pigs and monkeys. The last documented report of these large parrots was in 1680 by which time they were near extinction. I've always been fond of parrots and seeing a life-size statue of this extremely large parrot, perhaps the largest parrot that ever lived, was a revelation. The Large Raven Parrot is proof that Mauritius was once home to creatures every bit as unusual as the dodo yet it is only the dodo that is a celebrity of extinction.

Mauritius was also home to other members of the parrot family but today just one remains, the Echo Parakeet (*Psittacula eques echo*). Only through heroic conservation efforts has this species been saved from extinction. During the 1980s, only about 10 of these birds survived in the wild. Today, the Echo Parakeet is still considered critically endangered but its numbers have increased into the hundreds. This striking bird is emerald green with a black ring around its neck and blue-tinged feathers on its nape and tail. The Echo Parakeet is the sole surviving species of parakeet in the Mascarene Islands. Other species and subspecies on La Réunion and Rodrigues Island have become extinct.

Given the history of environmental damage inflicted on Mauritius by humans, it's easy to be pessimistic about the prospects for the success of conservation programs in the country. Islands have fragile ecosystems easily threatened by introduced species. Also, wildlife on Mauritius has the added burden of sharing a small land area of around 700 square miles with a relatively large human population of around 1.3 million.

On the other hand, there is sufficient evidence for cautious optimism. Mauritius is one of the few places on earth that has not only saved important endemic species from the absolute brink of extinction, but also has developed programs to protect them and greatly increased their numbers. Many countries around the world face similar problems and may look to Mauritius as an example of a developing nation that has pioneered successful environmental conservation strategies.

Only time will determine the conservation programs' success. No one is expecting to resurrect the dodo but if Mauritius can maintain a healthy environment for the remaining endemic species, who knows what will evolve.

CHAPTER 12

DODO LEGACIES

MAURITIUS IS A SURPRISING PLACE in many ways but especially in its natural beauty and the cultural tapestry of races and religions. Its environment and the species produced in this natural laboratory are unique in the world, and its human history is equally unusual. A truly multicultural society, Mauritius is a stable democracy with the ability to leverage historic cultural ties with Africa, India, China, and Europe.

Mauritius for me was a personal discovery. It is rarely on the radar for American tourists and not a place I thought I would come back to time and time again. I just happened upon this unfamiliar land because it was a convenient layover. In that sense, I am no different

45. Illustration of a dodo from 1860

from the Portuguese, Dutch, and French explorers who came here starting in the sixteenth century. For most of its history Mauritius was just a stop along the way, but then revealed itself to be so much more.

The dodo, Mauritius' most famous inhabitant, may be gone but the environs that created such an eccentric bird survives. I was amazed that one-of-a-kind species such as the Mauritian Kestrel and Pink Pigeon hung on as long as they did. Though it was humans that brought them to the brink of extinction, it was human intervention that saved these endangered birds. And I was impressed by the scope of conservation efforts to keep these and other species alive. Maybe by preserving the habitat and the remaining endemic flora and fauna on Mauritius, we can create a window into understanding our past and preserving our future.

I was surprised to learn there is very little scientific knowledge surrounding the dodo. The dodo is perhaps the most famous bird in history and at the same time one of the least understood. It only made a brief appearance to humans, and left little in the way of bones or fossils. Yet it became the very icon of extinction.

Today's successful conservation and environmental restoration programs in Mauritius arrived more than 300 years too late to save the dodo but are helping to reverse centuries of neglect, ignorance, and exploitation. The dodo's legacy inspired conservation programs because its extinction was the first connected to human activity. Now the battle is underway to prevent the surviving wildlife of Mauritius from going down the dodo path. In many ways it is a microcosm of the challenges facing the entire planet.

As residents of a small island nation, Mauritians are keenly aware of the threat of global warming and have already noticed its effects. Weather patterns have changed, altering the patterns of storms including cyclones, and periods of drought have become more

46. Rempart Mountain, Mauritius

common. Low-lying island nations are among the most threatened by rising sea levels. In response to climate change, both the government and private sector of Mauritius have embarked on plans for energy self-sufficiency including solar and wind technology, and are working to improve water management and local food production

Mauritius also faces great economic challenges. Plans are underway to grow the economy by doubling the number of tourist arrivals from around one million annually to two million within five years. This will place strains on infrastructure and the local environment, which must accommodate a growing population.

With growth comes economic opportunities however, and I think most people are aware of the urgent need to preserve the environment. Without its beautiful beaches, pristine waters, and incongruous vegetation, Mauritius would not have a tourist industry. And tourism has been the country's growth engine since its independence in 1968.

With little in the way of natural resources or land, Mauritius has leveraged its human resources to create one of the most dynamic economies in the developing world. I expect it will continue to do so in the twenty-first century, aided by new technology and an increasingly well-educated populace. Mauritius holds the potential to become a major international economic hub in the Indian Ocean.

And as the island grows with an eclectic population all living in close proximity, differences and tensions are bound to erupt. Yet I found the vast majority of Mauritians think of themselves first and foremost as Mauritians and I don't expect this will change. There is a strong sense of Mauritian national identity throughout the country.

Supposedly Mauritius is one of the most densely populated countries in the world, and if you only visit Port Louis or the crowded streets of the larger cities and towns, you might think that it is indeed like Hong

47. Engraving of a dodo from 1785

Kong. But visitors who venture out to explore the rest of the island, especially the south and east, will see vast fields of sugarcane, wide sandy beaches, and national forests that make it feel like you are on a very remote and empty island.

I learned more about nearby La Réunion as I explored Mauritius and became aware of the deep historical ties that link these two islands. La Réunion is similar in terms of geographic size and geology, and faces many of the same challenges that confront a small island with a fragile ecosystem and growing population.

Politically the two islands are very different countries facing their own specific issues. Modern La Réunion is an extension of mainland France floating in the Indian Ocean half a world away from Paris. Financial support from France and the European Union make it one of the wealthiest islands in the Indian Ocean. Mauritius is an independent sovereign country that must forge and fund its own economic path without much in the way of natural resources and has created one of the most dynamic economies in the developing world largely by leveraging its human capital since independence in 1968.

Of the three main Mascarene Islands, Rodrigues remains the most unspoiled and undeveloped. A visit to Rodrigues feels like a step back in time to a quieter, simpler era. It won't be easy, but I hope that the lessons of the dodo will continue to guide these Mascarene Islands as they attempt to balance the needs of a growing population with the need to preserve some of the most unique and irreplaceable natural environments on earth.

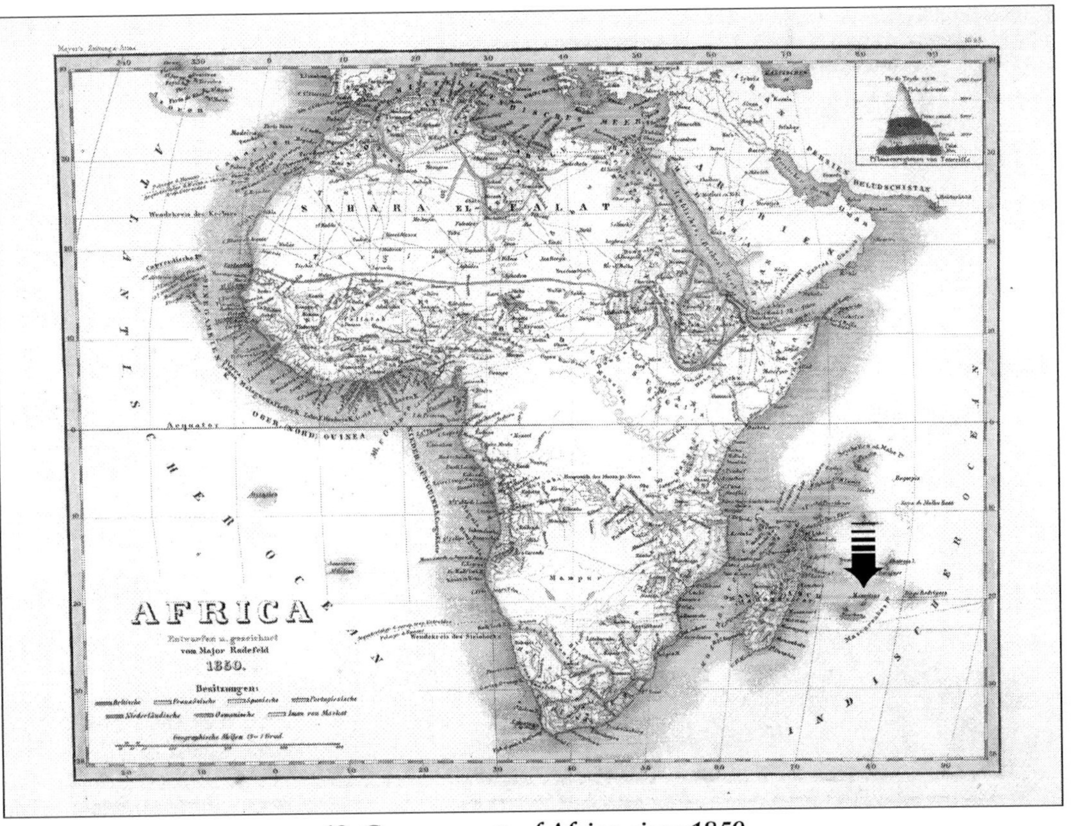

48. German map of Africa circa 1850

REFERENCES

1. Carroll, Lewis, *Alice's Adventures in Wonderland* (Oxford, 1865).
2. Cheke, Anthony and Hume, Julian P., *Lost Land of the Dodo: The Ecological History of Mauritius, Reunion, and Rodrigues,* 1st ed. (Yale University Press, 2008).
3. Corn, Charles, *The Scents of Eden* (Kodansha International, 1998)
4. De Saint-Pierre, Bernadin and Wilson, Jason, *Journey to Mauritius* (Interlink Books, 2003).
5. Durrell, Gerald, *Golden Bats and Pink Pigeons* (House of Stratus, 2002).
6. Flannery, Tim, *A Gap In Nature: Discovering the World's Extinct Animals* (Atlantic Monthly Press, 2001).
7. Fuller, Errol, *Dodo: A Brief History* (Universe Publishing, 2003).
8. Fuller, Errol, *Extinct Birds* (Cornell University Press, 2001).
9. Grihault, Alan, *Dodo: The Bird Behind the Legend* (Imprimerie & Papeterie Commerciale, IPC, 2005).
10. Grihault, Alan, Solitaire: *The Dodo of Rodrigues Island* (Precigraph Ltd., 2007).
11. Hachisuka, Masauji, *The Dodo and Kindred Birds* (H. F. & G. Witherby, 1953).
12. Le Comte, Christian Carlos Guillermo, *Mauritius from Its Origin* (Christian le Comte, 2005).
13. Leguat, François, *The Voyage of François Leguat of Bresse to Rodriguez, Mauritius, Java, and the Cape of Good Hope:* Transcribed from the First English Edition (Cambridge University Press reissue, 2010)

14. Morgan, Helen, *Blue Mauritius: The Hunt for the World's Most Valuable Stamps* (Atlantic Books, 2008).

15. The Mauritius Collection at the National Library of Australia, Canberra.

16. Pearson, Michael, *The Indian Ocean* (Routledge, 2003)

17. Piat, Denis, *Mauritius on the Spice Route 1598–1810* (Editions Didier Millet Pte Ltd).

18. Piat, Denis, *Pirates & Corsairs in Mauritius* (Christian le Comte, 2007).

19. Pinto-Correia, Clara, *Return of the Crazy Bird: The Sad, Strange Tale of the Dodo* (Springer, 2003).

20. Quammen, David, *Song of the Dodo: Island Biogeography in an Age of Extinctions* (Scribner, 1996).

21. Saint-Pierre, Bernardin De, *Journey to Mauritius* (Interlink Books, 2002).

22. Saint-Pierre, Bernardin De, *Paul & Virginia* (Peter Owen Publishers, 2005).

23. Turner, Jack, *Spice: The History of a Temptation* (Vintage Books, 2005)

ILLUSTRATIONS

Figure 29: "Habitation des Pamplemousses Jardin de M. Cere." Steel engraving by Lemaitre, Paris, circa 1840.

Figure 30: Steel engraving of Pierre Poivre (1719-1786) by E. Conquy, France, 1835.

Figure 31: Spice plants: nutmeg, cinnamon, clove and allspice by Blackie & Son., Glasgow, Edinburgh, & London, circa 1866.

Figure 32: 1780 Map of Réunion, Mauritius, and Rodrigues Island: "Carte Générale Des Isles dont les figures ci-dessus offrent un plus ample detail". Hand-colored copper etching by Rigobert Bonne (1729–1795) for "Atlas de toutes les parties connues du globe terrestre" by Guillaume Thomas François Raynal (1713–1796). Genève, Switzerland, 1780.

Figure 35: Leguat Solitaire, from the *Pictorial Museum of Animated Nature,* 1844.

Figure 38: Didus Ineptus, hand-colored engraving from *General Zoology,* London, dated 1806.

Figure 39: "Dodo Presenting a Thimble" by Sir John Tenniel from *Alice's Adventures in Wonderland* by Lewis Carroll, 1865.

Figure 40: Dodo skull and foot, from the *Pictorial Museum of Animated Nature,* 1844.

Figure 41: "Mouse telling story to birds and Alice" by Sir John Tenniel from *Alice's Adventures in Wonderland* by Lewis Carroll, 1865.

Figure 42: *Dodo, Cassowary Ostrich, and Penguins* by F. J. Bertuch. Hand-colored copper engraving for Builderbuch fur Kinder, Germany, 1790.

Figure 43: Plate of a dodo from *A History of the Earth & Animated Nature* by Oliver Goldsmith, London, 1790.

Figure 45: "The Dodo" from Cassell's *Popular Natural History,* London, circa 1860.

Figure 47: Engraving of dodo by William Frederic Martyn, London, 1785.

Figure 48: German map of Africa, *Meyer's Zeitung-Atlas,* 1850

PHOTOGRAPHS

All photographs listed below are the work of the author:

Figure 33: Saint Gabriel Church, Rodrigues Island.

Figure 34: Tethered cattle are common sight on Rodrigues.

Figure 36: The prison on Rodrigues Island is decorated with a Solitaire.

Figure 37: One of many beautiful beaches on Rodrigues Island.

Figure 44: View of Mahébourg Bay from Île aux Aigrettes.

Figure 46: Rempart Mountain, Mauritius.

ABOUT THE AUTHOR

The author in Mauritius, 2011

Tom Parker is a writer based in Seattle, Washington and has written professionally for the software, wine, and other industries. His lifelong interest in history and travel inevitably led him to write on these subjects. He documents journeys with his own photographs and is continually adding new destinations to his travel plans with a focus on noteworthy, if esoteric subjects that capture his imagination.

INDEX

12830332R00111

Made in the USA
Charleston, SC
31 May 2012